AF598262

Boundary Elements in Fluid Dynamics

INTERNATIONAL CONFERENCE ON COMPUTER MODELLING OF SEAS AND COASTAL REGIONS AND BOUNDARY ELEMENTS AND FLUID DYNAMICS
SOUTHAMPTON, U.K., APRIL 1992

International Scientific Advisory Committee for Boundary Elements and Fluid Flow

A. Alujevic
C.A. Brebbia
M. Bush
J. Connor
G. De Mey
U. Gulcat
H. Hu
K. Kitagawa
P.W. Partridge
H. Power
H. Schmidt
R. Shaw
M. Tanaka
L.C. Wrobel
J. Wu

Acknowledgement is made to K. Kitagawa *et al.* for use of Fig. 11, page 19, which appears on the front cover of this book.

Boundary Elements in Fluid Dynamics

Editors: C.A. Brebbia, Wessex Institute of Technology
P.W. Partridge, Wessex Institute of Technology

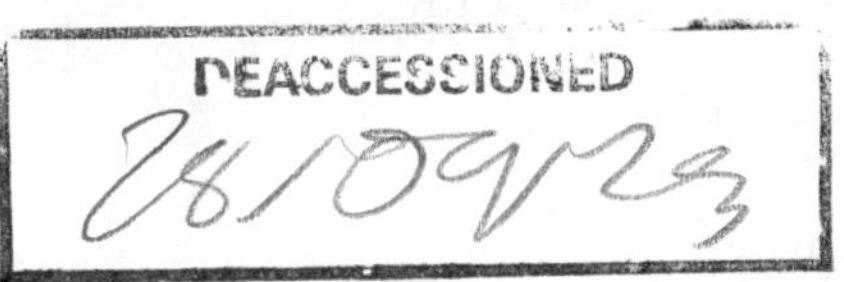

Computational Mechanics Publications
Southampton Boston

Co-published by

Elsevier Applied Science
London New York

P.W. Partridge
Wessex Institute of Technology
Ashurst Lodge
Southampton
SO4 2AA
U.K.
(also from the Dept. of Civil Engineering
University of Brasilia, Brazil)

C.A. Brebbia
Wessex Institute of Technology
Ashurst Lodge
Southampton
SO4 2AA
U.K.

Co-published by:

Computational Mechanics Publications
Ashurst Lodge, Ashurst, Southampton, UK

Computational Mechanics Inc.
25 Bridge Street, Billerica, MA 01821, USA

and

Elsevier Science Publishers Ltd
Crown House, Linton Road, Barking, Essex IG11 8JU, UK

Elsevier's Sole Distributor in the USA and Canada:

Elsevier Science Publishing Company Inc.
655 Avenue of Americas, New York, NY 10010, USA

British Library Cataloguing in Publication Data
A Catalogue record for this book is available
from the British Library

ISBN 1-85166-780-6 Elsevier Applied Science, London, New York
ISBN 1-85312-165-7 Computational Mechanics Publications, Southampton
ISBN 1-56252-093-8 Computational Mechanics Publications, Boston, USA

Set
ISBN 1-85166-799-7 Elsevier Applied Science, London, New York
ISBN 1-85312-193-2 Computational Mechanics Publications, Southampton
ISBN 1-56252-121-7 Computational Mechanics Publications, Boston, USA

Library of Congress Catalog Card Number 91-77631

No responsibility is assumed by the Publishers for any injury and/or damage to persons or property as a matter of products liability, negligence or otherwise, or from any use or operation of any methods, products, instructions or ideas contained in the material herein.

©Computational Mechanics Publications 1992

Printed and bound by The Alden Press, Ltd., Oxford

All rights reserved. No part of this publication may be reproduced, stored in a retrieval system or transmitted in any form or by any means, electronic, mechanical, photocopying, recording, or otherwise, without the prior written permission of the publisher.

CONTENTS

SECTION 1: TURBOMACHINERY

SECTION 2: AERODYNAMICS

SECTION 3: VISCOUS FLOW AND TURBULENCE MODELS

PREFACE

This book *Boundary Elements in Fluid Dynamics* is the second volume of the two volume proceedings of the International Conference on Computer Modelling of Seas and Coastal Regions and Boundary Elements and Fluid Dynamics, held in Southampton, U.K., in April 1992.

The Boundary Element Method (BEM) is now fully established as an accurate and successful technique for solving engineering problems in a wide range of fields.

The success of the method is due to its advantages in data reduction, as only the boundary of the region is modelled. Thus moving boundaries may be more easily handled, which is not the case if domain methods are used. In addition, the method is easily able to model regions to extending to infinity.

Fluid mechanics is traditionally one of the most challenging areas of engineering, the simulation of fluid motion, particularly in three dimensions, is always a serious test for any numerical method, and is an area in which BEM analysis may be used taking full advantage of its special characteristics.

The conference includes sections on turbomachinery, aerodynamics, viscous flow and turbulence models, and special flow situations. The organisers would like to thank the International Scientific Advisory Committee, the conference delegates and all of those who have actively supported the meeting.

C.A. Brebbia
P.W. Partridge
April 1992

SECTION 1: TURBOMACHINERY

A Numerical Simulation of Cross Flow Fan

K. Kitagawa (*), H. Tatsuke (**), Y. Tsujimoto (**), Y. Yoshida (**)
(*) Airconditioners & Appliances Eng. Lab., Toshiba Corp., 8, Shinsugita-cho, Isogo-ku, Yokohama, 235, Japan
(**) Mechanical Engineering, Faculty of Engineering Science, Osaka University, Toyonaka, Osaka, 560, Japan

ABSTRACT

The singularity method, a kind of boundary element method, is applied to a cross flow fan, which is widely used for household separate types of air-conditioners. A potential flow analysis is carried out for the cross flow fan with a complicated shape of casing. The flow is modeled by bound source/vortex distributions on the boundaries and blades, and free vortices shed from the impeller blades following Kelvin's circulation theorem. Numerical simulation results are in good agreement with experimental results.

INTRODUCTION

It is in the late 19th century that the cross flow fan was developed from the ideas of Mortier. After Eck[1,2] and Laing[3] devoted considerable effort to the cross flow fan in the early 1950s, it was applied to practical units. Recently it has been widely used for small domestic appliances, in particular, household separate types of air-conditioners.

The cross flow fan has an impeller of drum type, and air enters along the full width of the fan, perpendicular to the axis of the rotation. These characteristics are similar to the sirocco fan, however, the cross flow fan is closed at both ends and can be designed longer in the axial direction. It is complicated and difficult to design the cross flow fan. The flow created by the cross flow fan is characterized by the formation of a line vortex eccentric to the axis of rotation. In addition, the flow field is affected by shapes of both the impeller and the casing. Because of difficulty in theoretical analyses, the cross flow fan has been generally developed by experimental research.

In order to improve the performance of the cross flow fan, numerical simulation approaches are also important. Not so many papers were available in this problem, in spite of actual needs. Ikegami and Murata(4) and Yamafuji(5) developed potential flow analyses by applying an actuator disk theory to the impeller with a simple shape of casing. Quite recently for more realistic and complicated shapes of casing, Iizuka et al(6) applied a Clouds-in-Cells method combined with the FEM and Okamoto et al(7) applied a discrete vortex method with a singularity method.

In this paper, we also develop an inviscid flow analysis by using the discrete vortex method combined with the singularity method. The flow is modeled by bound source/vortex distributions on the boundaries and blades, and free vortices shed from the impeller blades following Kelvin's circulation theorem. In the problem of the complicated shape of casing, the proposed numerical approach is compared with experimental results.

NOMENCLATURE

NB	:	Number of blades on the impeller
NC	:	Number of unknown points on the backside of casing
ND	:	Number of unknown points on the outlet
NS	:	Number of unknown points on the inlet
NT	:	Number of unknown points on the frontside of casing
dt	:	Time increment
r	:	Radius
q	:	Source distribution
u,v	:	x, y direction components of absolute velocity
α	:	Dissipation coefficient for shed vortex
β	:	Blade angle
γ	:	Vortex distribution on the boundary and the blade
ϕ	:	Flow rate coefficient
φ	:	Pressure coefficient
ω	:	Vortex distribution in the domain
Ω	:	Angular velocity of the impeller

Subscripts

1,2	:	Outer and inner periphery of impeller respectively
r,θ	:	Radial and peripheral components respectively
t,tk	:	All and region excluding the eccentric vortex of impeller respectively.

THEORY AND NUMERICAL IMPLEMENTATION

In this paper, a two-dimensional potential flow analysis is applied and fluid is assumed incompressible and inviscous.

By applying Green's theorem in a complex variable space, a complex velocity, u-iv, is expressed in Eq.(1) inside of a domain F enclosed by a boundary L.

$$u-iv = \frac{1}{2\pi}\int_L \frac{q(s)+i\gamma(s)}{Z-Z(s)}ds + \frac{1}{2\pi}\iint_F \frac{q_F(x)+i\omega_F(x)}{Z-Z(x)}dF \quad (1)$$

$q(s)$: Source distribution on the boundary L
$\gamma(s)$: Vortex distribution on the boundary L
$q_F(s)$: Source distribution in the domain F
$\omega_F(s)$: Vortex distribution in the domain F

In this paper, a shape of casing is assumed to be shown in Fig. 1. From Cauchy's integral representation, the boundaries and blades can be represented by source-vortex distributions, q (s) and $\gamma(s)$, which are equivalent to the normal and tangential velocity components on the boundary, respectively. Then, we can model the casing with distributions of unknown q(s) and $\gamma(s)$ as indicated in Fig. 1. The impeller is expressed as unknown vorticity distributions in finite points of each blade, which rotates with time. As a second term of Eq. (1), only distributions of vorticity $\omega_F(x)$ need to be evaluated as an effect of vorticities shed from each blades. These variables are made dimensionless by choosing r_1: an outer radius of the impeller and $r_1\Omega$: a peripheral component of velocity in the outer radius as length and velocity, respectively.

Boundary conditions applied are as follows: On the inlet of casing, uniform distributions of normal velocity are given. On the outlet of casing, discharged flow direction is assumed to be normal. On the rest boundary of casing, normal velocity is assumed to be zero. On the blade surface, normal components of relative velocity are equaled to be zero and unsteady Kutta's condition is applied at the trailing edges of each blade. As a result, the strength of vorticities shed from the impeller is determined by Kelvin's circulation theorem.

The shed free vortex is assumed as a point vortex with a small radius r_s. In order to approximate the effects of viscous or turbulent diffusion of the vorticity, the strength of the shed vortices are simply reduced with time as expressed in Eq.(2).

$$\omega_F^{(n+1)} = \omega_F^{(n)} \times e^{-i\alpha\Delta t} \tag{2}$$

A flow chart of proposed method is shown in Fig. 2. An initial flow field is obtained by carrying out a simple potential analysis for the casing which has no impeller. With an increment dt on time, the impeller rotated and new shed vorticities are created at the trailing edge of each blade. To reduce computation time and avoid disconvergence caused by an interaction of shed vorticities situated in a quite small distance, near-by vortices are merged. Numerical computation is continued, until flow fields become stable.

COMPUTATIONAL RESULTS

In this section, several numerical results, including the comparison with experimental results, obtained by the present analysis method are discussed. Fig. 3 shows a shape of casing discussed and numbers of discretization used for boundary. The impeller consists of NB blades and each blade is modeled by NM points of vortex. Circular arc vanes with inner and outer vane angles β_1 and β_2 are used in the present examples.

Fig. 5 is the time plots of the location of the center of the eccentric vortex and the number of shed vortices, showing the process of convergence. Parameters used for computation are as follows: time increment dt=0.05, dissipation coefficient for shed vortices α=0.3, flow rate coefficient ϕ=0.5, leading and trailing angles of the blade β_1, β_2=20, 100. After the time step reaches about 250, that is, the impeller rotates two times, the flow field computed could be judged almost converged. The velocity vectors and distributions of shed vortices, obtained when the time step equals to 350, are shown in Fig. 6. Black and white circles mean shed vortices of clockwise and counterclockwise directions, respectively. An initial flow field, which is obtained without the impeller, is shown in Fig. 7. It should be noted that there is no eccentric vortex in this result. However, after certain time steps, the counter-clockwise free vortices shed from the blades are accumulated and forms an eccentric vortex. Numerical computation was carried out by using TITAN3000 and CPU time was about 4 hours. The most time consuming routine was a transfer process of shed vortices.

Next, numerical results are compared with experimental results to confirm validity of the proposed analysis. An experimental flow field obtained by flow visualization in the water model, is shown in Fig. 8. By comparing Figs. 7 and 8, the flow fields, in particular, locations and patterns of the eccentric vortex, are in good agreement.

We also investigate a method for evaluating visualized flow fields by using a computer image processing. In Fig. 9, velocity distributions obtained by the above experimental method and the present analysis are compared inside and just outside of the impeller. From a theoretical reason, the experimental process has a difficulty in evaluating unstable or turbulent velocity distributions. That is a little less accurate near the eccentric vortex situated in about θ=320. Taking this point into consideration, these results are in quite good agreement.

At last, results computed at ϕ=0.7, 0.3 are discussed. Flow fields described by velocity vectors and distributions of shed free vortices are shown in Figs. 10 and 11, respectively. There is almost no significant difference between results of ϕ=0.7 and 0.5, however, the result obtained by ϕ=0.3 slightly differs from both. That is, a small amount of shed vortices accumulate at the inlet near the backside casing and the location of eccentric vortex moves to $\theta=300^\circ$ from $\theta=320^\circ$.

Table 1 shows the head coefficients for three different flow coefficients obtained from the balance of angular momentum, which can be used as a guideline for design. As shown in Fig. 5, flow fields are slightly oscillating even after a certain transient period at start up. This suggests another possibility of the application of the present simulation; noise evaluation might be possible from the force fluctuation on the blades.

CONCLUSION

In this paper, numerical analysis method for flow fields created by the cross flow fan are developed. An inviscid flow analysis is carried out by using the singularity method and the discrete vortex method. With the Cauchy's integral representation, the cross flow fan is modeled by vorticity distributions on the boundaries and blades, and free vorticities shed from the impeller. Computational results are compared with experimental results and it is shown that they are in good agreement.

REFERENCES

(1) ECK, B. Cross-flow blower, Brit. Pat. 757,543, 1954.

(2) ECK, B. Ventilatoren (Fans). Berlin: Springer Verlag, 1962.

(3) LAING, N. Cross-flow fan. Brit. Pat. 943,765, 1959.

(4) IKEGAMI, H. MURATA, S.: A Study of Cross Flow Fan. Technology Rep. Osaka Univ. 557, 16, 1966.

(5) YAMAFUJI, K: Proceeding of the JSME, Vol. 41, No. 344, 1184, 1975 (in Japanese).

(6) IIZUKA, et al: Proceeding of the JSME, No. 914-1, 1991 (in Japanese)

(7) OKAMOTO Y. et al: Discrete Vortex Method Analysis of Cross Flow Fan, Proceeding of the JSME, No. 917-1, 170, 1991 (in Japanese).

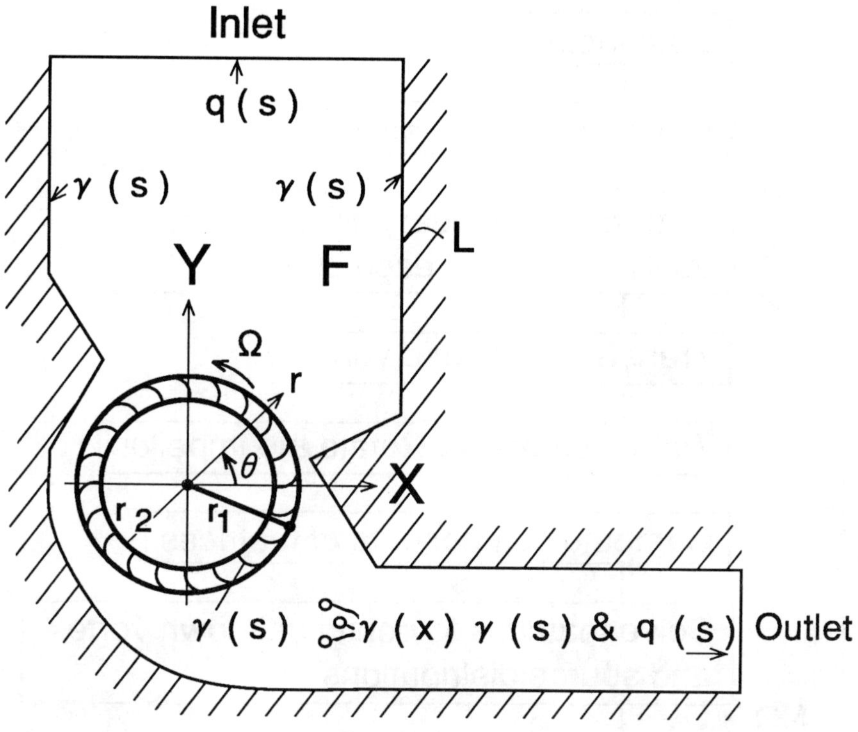

Fig. 1 Analytical model

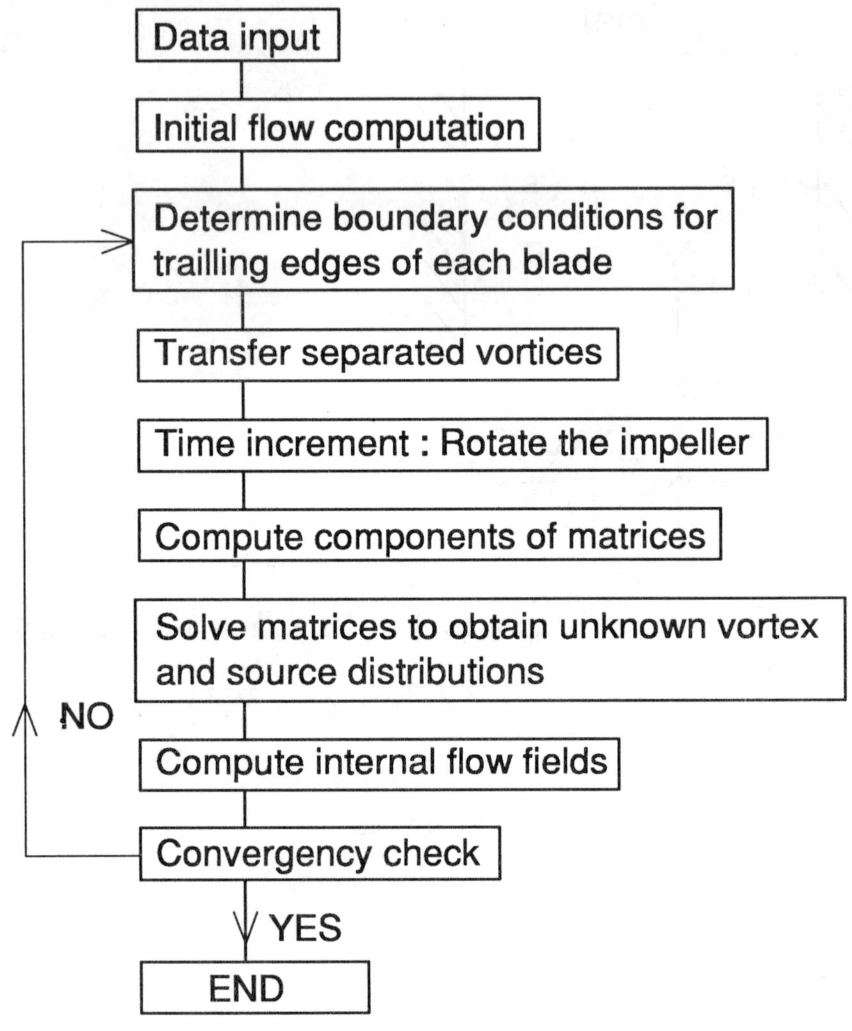

Fig. 2 Flow chart of the present analysis

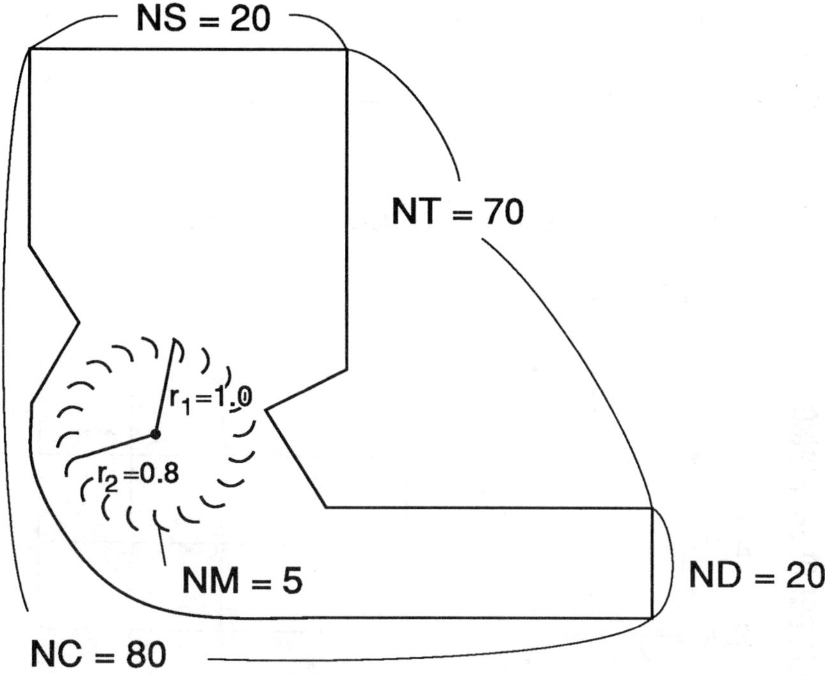

Fig. 3 Computational model and discretization

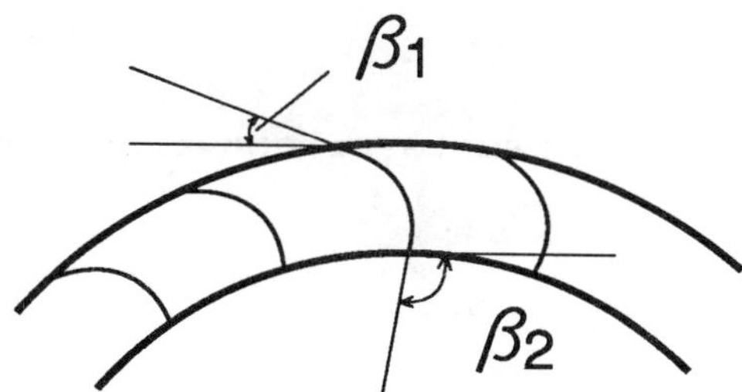

Fig. 4 Impeller and blade model

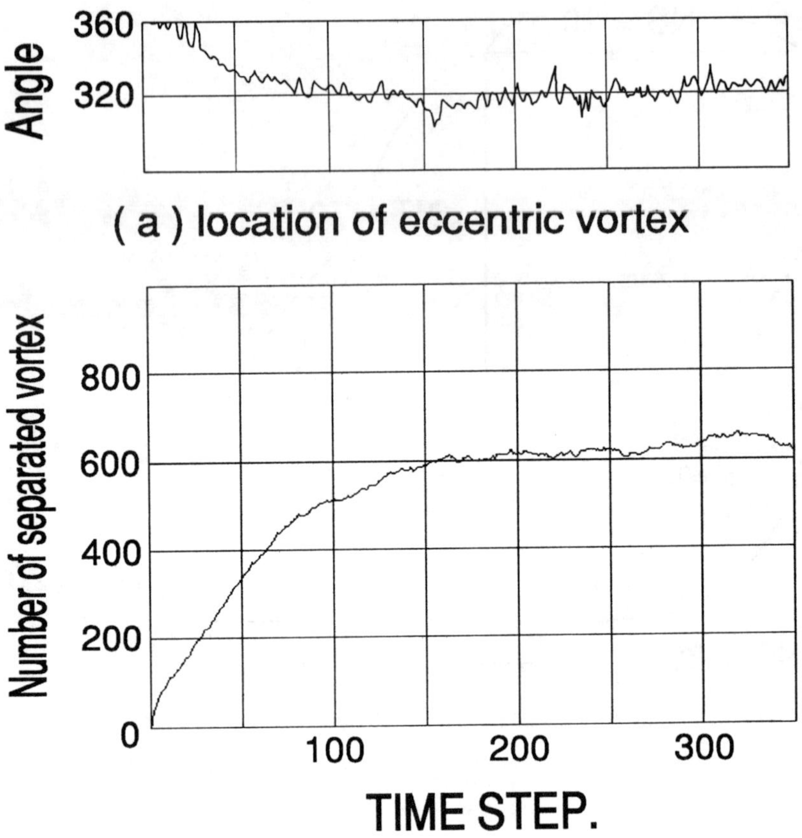

Fig. 5 Computation results of convergent process

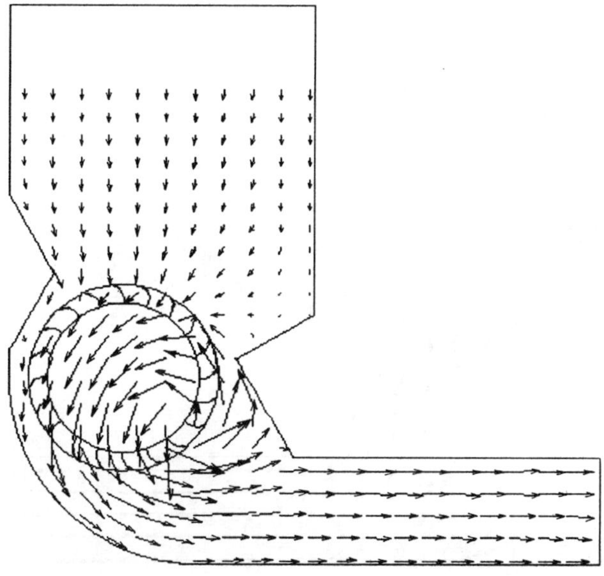

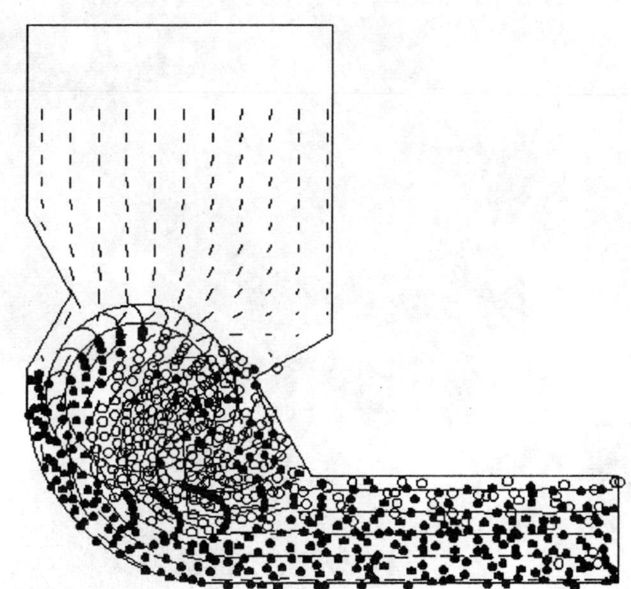

Fig. 6 Computation results of flow fields and shed vortices

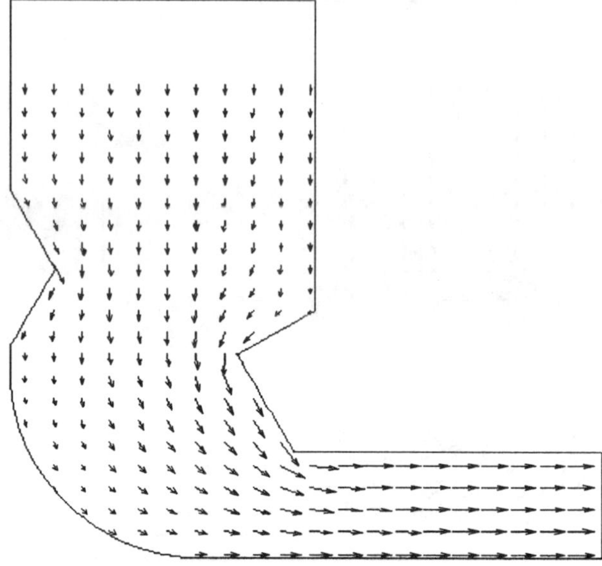

Fig. 7 Initial flow fields

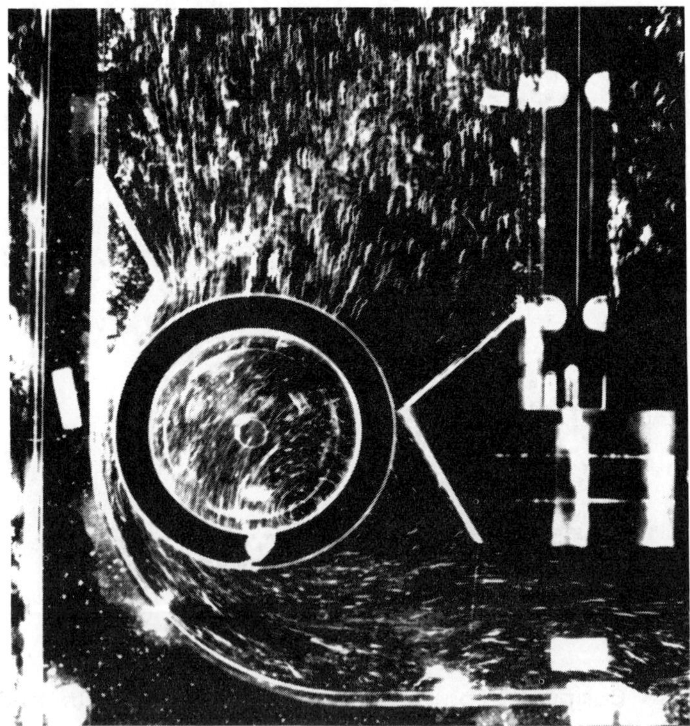

Fig. 8 Flow fields obtained by experimental visualization

○ Experimental V r ——— Analysis result

● Experimental V θ

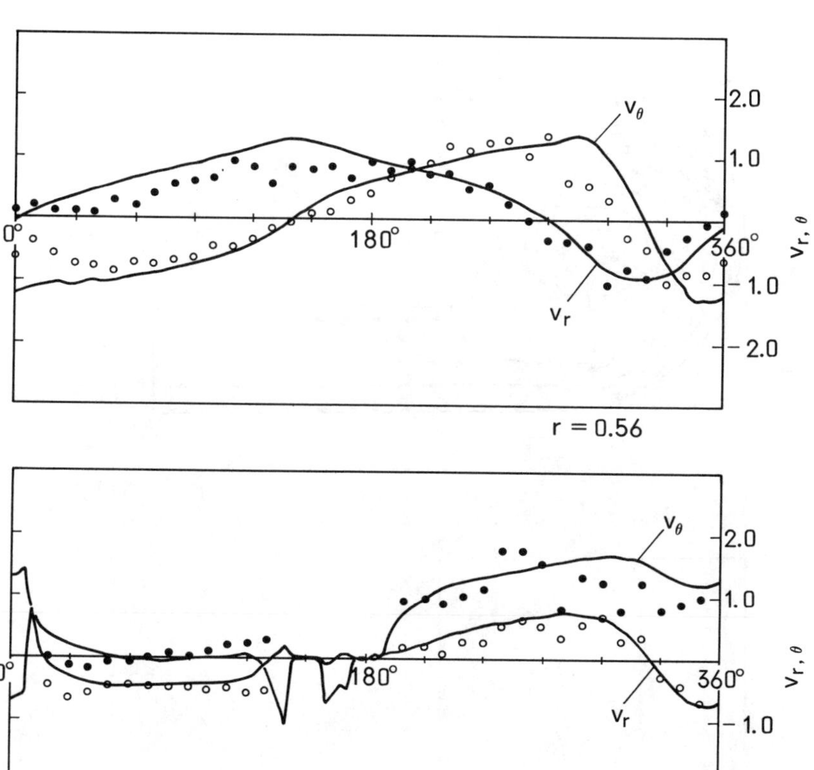

Fig. 9 Comparison of velocity distributions

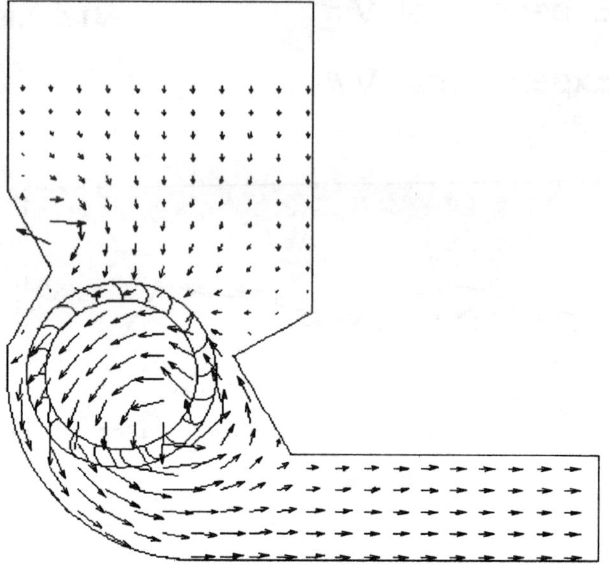

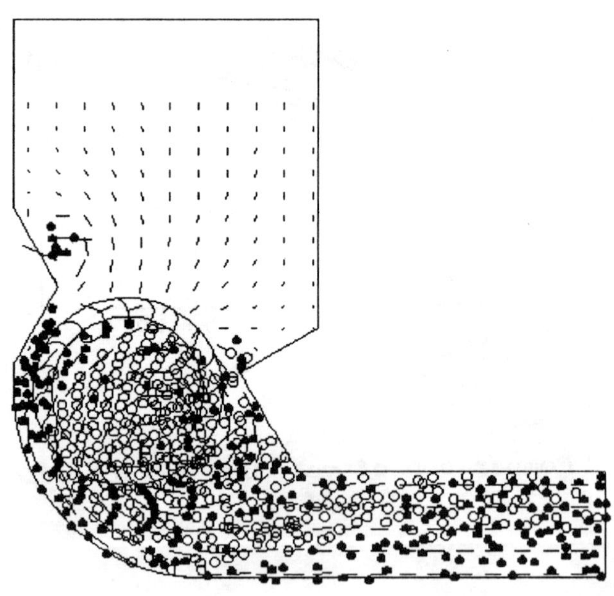

Fig. 10 Computed flow fields (ϕ=0.3)

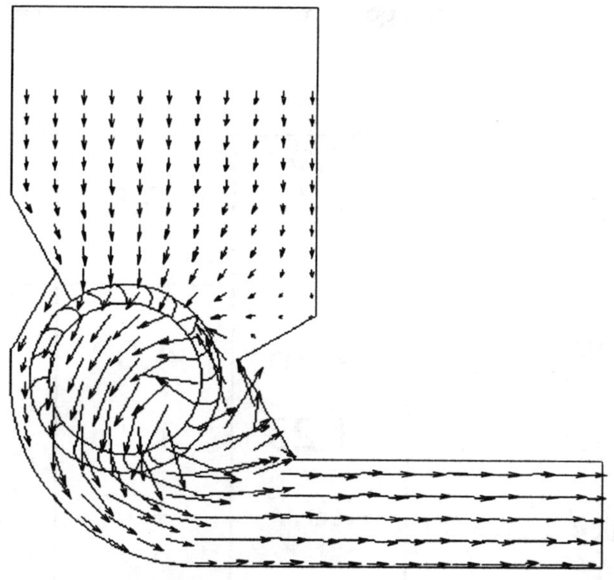

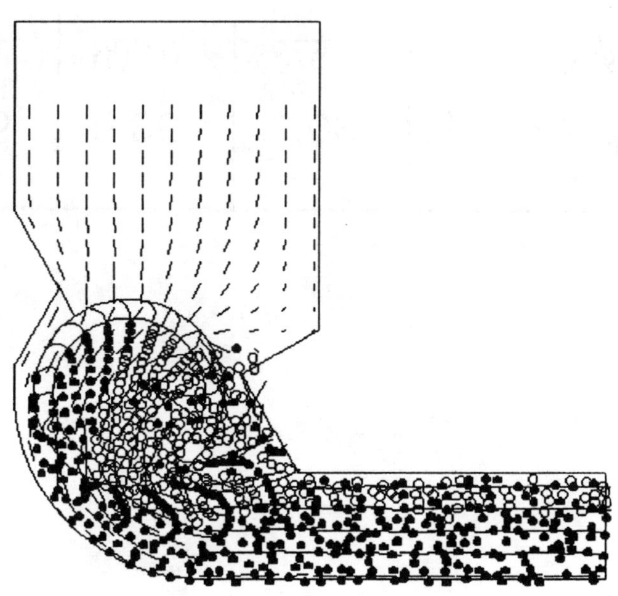

Fig. 11 Computed flow fields (ϕ=0.7)

LUCY CAVENDISH COLLEGE

Table 1 Computed performance characteristics

Φ	0.3	0.5	0.7
Ψ_t	4.46	4.03	4.86
Ψ_{tk}	1.28	1.76	1.94
Ψ_{tk} / Ψ_t (%)	28.7	43.7	39.9
ϕ_t	0.885	1.08	1.33
ϕ_{tk}	0.573	0.906	1.21
ϕ_{tk} / ϕ_t (%)	60.7	83.9	91.0

Fully 3D Euler Solutions Versus Test Results for the Whole Load Range of Radial Pumps

J. Riedler (*), E. Goede (**)

(*) Sulzer Bros., Winterthur, Switzerland

(**) Sulzer Escher Wyss, Zurich, Switzerland

Abstract

16 different impellers of the specific speed range from 17 to 33 have been analysed by means of a fully 3-D Euler code at the best efficiency point. Comparisons with the theoretical head derived from test results have yielded a nearly 50% reduction against the scatter band width of empirical correlation methods.

Part load flow patterns have been calculated for the range between shut off and the best efficiency point. Realistic velocity profiles have been obtained even for recirculating flow.

Introduction

A 3-D Euler-code developed at the Ecole Polytechnique de Lausanne has become a powerful design tool for hydraulic turbines at Sulzer Escher Wyss. It has already been successfully applied for pump turbines. Based on this program a special pump code has been developed which has been in use since 1988. Detailed information can be obtained from the reports of Goede et al. [1,3,4] and from Thibaud et al. [2].

A large number of impellers have been investigated by applying the fully 3-D Euler code and test results are already available for comparisons. Most of the impellers have been tested with radial inlet, some with axial inlet and only one with both inlet devices.

It was found that the Euler-code works also at part load. Therefore calculations have been performed for the whole load range.

Normally, impeller recirculation is considered as a purely viscous phenomenon. Therefore it was somewhat surprising that realistic velocity profiles for recirculating flow at the impeller inlet can also be derived from an inviscid fully 3-D Euler code.

In contrast to the potential flow formulation, the Euler approach accounts for the rotational effects within the flow. Consequently, the motion of vortices and secondary flow can be predicted.

The finite volume technique which is used in the Euler-code, requires the conservation laws in intergral form. For an incompressible flow, the Euler momentum equations for a rotating frame of reference:

$$\frac{\partial}{\partial t} \int_{Vol} \vec{w}\, dVol + \int_{S} \left[(\vec{w}\,\vec{n})\, \vec{w} + \frac{p}{\rho} n \right] dS = \int_{Vol} \vec{f}_B \, dVol \qquad (1)$$

must be solved along with the continuity equation:

$$\int_{Vol} (\vec{w}\,\vec{n})\, dS = 0 \quad . \qquad (2)$$

The equations contain the pressure p, the density ρ, the relative velocity $\vec{w}$, the normal vector $\vec{n}$ on the finite volume surfaces and the body forces $\vec{f}_B$ which include centripetal and Coriolis forces as well.

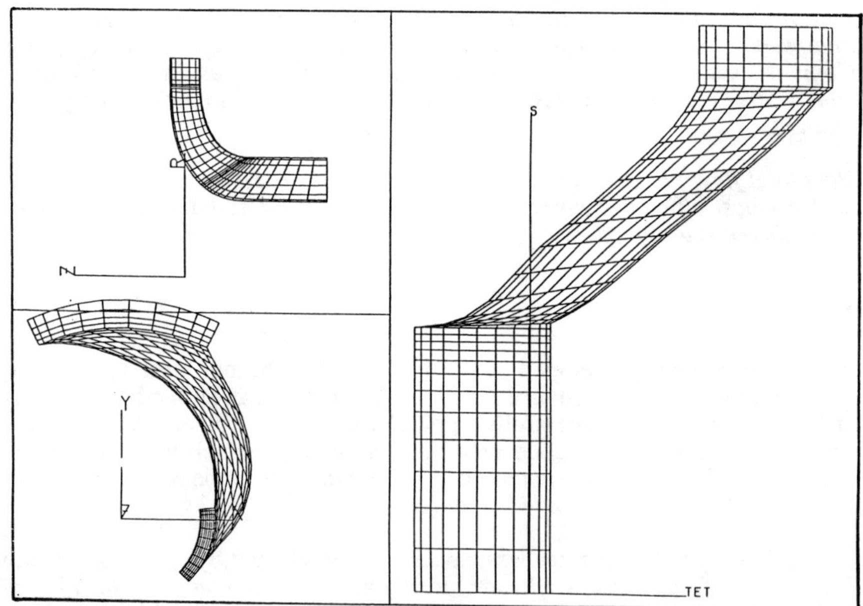

Fig. 1: Typical mesh used for 3-D Euler flow calculation

Unfortunately, the efficient time marching method as used by Thibaud, Drotz and Sottas [2], can only be applied if all equations are hyperbolic in time. If only the steady-state solution is of interest, an artificial pressure term has to be added to the continuity equation (2) as suggested by Chorin [5] and Peyret [6]. Therefore, the solution at a certain time step has no physical meaning unless the steady-state condition is reached. In addition, an artificial viscosity has to be introduced in order to achieve numerical stability.

For the discretization of the flow domain a socalled H-mesh was generated. A typical mesh used for the flow simulations is shown in fig. 1.

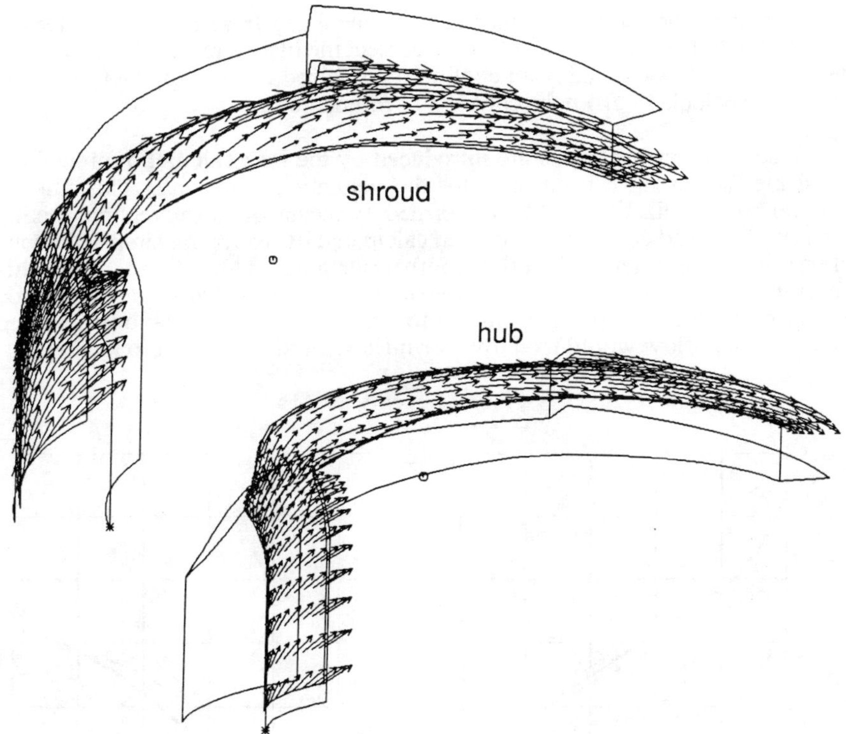

***Fig. 2**:* *Calculated velocity vectors for the best efficiency point*

The boundary conditions which have been applied are:

- the kinematic condition at solid walls (no through flow),
- prescribed velocity vectors at inlet according to discharge, and
- the static pressure at outlet.

In order to save computer time, only one blade channel has been considered, and therefore periodicity conditions have been used for the up-stream and downstream ducts of the impeller.

Head Prediction

The prediction of the accurate head produced by a pump is a difficult task. Even if the ideal head is considered, the various slip factor correlation methods produce a lot of scatter, because only a limited number of parameters will be taken into account. Therefore the 3-D Euler flow calculation is a major step towards more precise estimates, because it reflects all geometrical details. Limits arise only from the discretisation of the flow domain, from the stability of the numerical procedure, from the neglected viscous effects and from the boundary conditions which are not known very precisely during the design process.

The static pressure difference can be calculated from averaged values at inlet and outlet surfaces which do not intersect the bladed region. The dynamic head should be calculated from continuity by introducing shape factors which have to be calculated from the equation of momentum.

Due to the errors which are introduced by the flow calculation procedure, no shape factors have been used for the velocity components. This implies that the homogenization will be performed by means of an irreversible mixing process. The head coefficient has been calculated by averaging simply the total pressure at the inlet and outlet control surfaces. The advantage of this definition, which is only exact for the static pressure, is that it can be applied for the entire load range from shut off to over load, while a relationship based on the throughflow would tend towards infinity in the case of zero load.

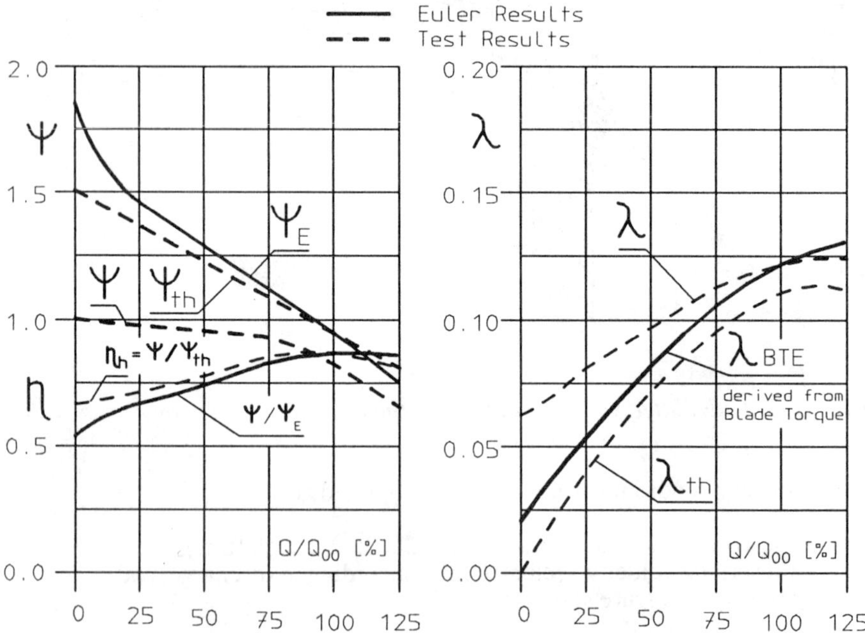

Fig. 3: Head prediction for pump operation (comparison with test results)

Fig. 4: Prediction of the power consumption (comparison with test results)

In fig. 3 the head coefficient ψ_E derived from the Euler flow calculation is shown along with the measured head coefficient ψ and the theoretical head coefficient ψ_{th}. For the whole load range the theoretical head coefficient was based on the slip factor, which was obtained from a loss analysis at the best efficiency point. The Euler results show that the slip factor increases slightly with a decreasing flow rate in the range of 40% to 125%. But as soon as recirculation starts at the impeller inlet a dramatic increase appears toward shut off when the flow rate will be reduced further. At zero load a head coefficient of 1.85 was derived, which is close to the expected value of 2. The discrepancy might be caused by the finite volume size and by the artificial viscosity.

The artificial viscosity together with the finite discretization of the flow domain and the extrapolation of the velocity to the blade surfaces are also responsible for the difference between the theoretical power coefficient λ_{th} and the power coefficient λ_{BTE} based on the blade torque obtained from the 3D Euler results. In the example of fig. 4, where the artificial viscosity was kept constant, the discrepancy is nearly the same for the investigated load range with only a slight increase at shut off and overload.

The measured power coefficient λ includes also disc friction, mechanical losses and recirculation losses and it seems that the recirculation losses are mainly responsible for the relatively large power consumption at shut off. In the ideal case no energy should be necessary for stationary operation at zero load if inlet and outlet are blocked. But due to the numerical viscosity there will always remain a small but finite blade torque at zero flow rate.

Slip factor correlation at the Best Efficiency Point

Different slip factor correlations have been applied. The best results have been obtained from a relationship based on empirical data. For impellers of normal design the scatter is very small. Strong deviations exist only in cases where the geometry was not designed according to standard procedures. In such cases the 3-D Euler flow calculation is superior, because it takes the whole geometry into account, while the slip factor correlation contains only few geometric parameters.

Quantitative comparisons at the Best Efficiency Point

Fig. 5 contains the predicted head of the 3-D Euler flow calculation and the slip factor correlation in function of the specific speed.The head refers to the theoretical head H_{th} derived from measurements at the operating point of best efficiency.

The slip factor correlation method yields an average value of 0.992 with a band width -8.1% to 10.3% for 21 tests carried out with 16 impellers. For the Euler results a mean value of 0.976 was obtained with a considerably smaller band width of -5.8% to 5.7%. The accuracy is pretty good, because the comparison also includes uncertainties introduced by reconstruction, manufacturing, loss analysis and boundary conditions.

An impeller with the specific speed of n_q=33 was tested as a single stage impeller with axial and radial inlet as well as a normal impeller in a multistage pump. The total band width due to different boundary conditions was approximately 6%. This is slightly more than half of the total band width obtained from all test results. Therefore, the fluctuation could probably be reduced further if the correct boundary conditions would be described. For the comparisons, the Euler results were calculated for uniform flow at the axial inlet.

Different mesh parameters have been used for the discretization of the flow domain as well as slightly different parameters for the artificial compressibility and the artifical viscosity, but it seems that their effect on the results is not significant.

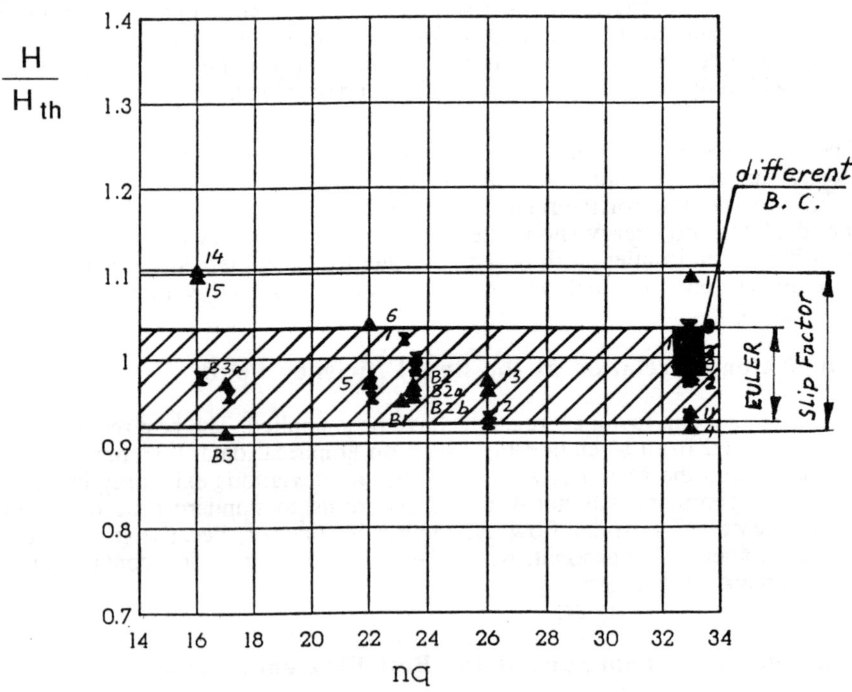

Fig. 5: *Band width for predicted head H with respect to theoretical head H_{th} as a function of the specific speed n_q (derived from 21 tests for 16 different impellers using 3-D Euler flow calculation and the slip factor method)*

Part load flow patterns

It is well known that impeller inlet recirculation starts at a certain part load flow rate at the impeller shroud (fig. 6). The flow leaves the impeller at the outer eye diameter and returns together with the approaching suction pipe flow by introducing some prerotation. This is the reason why the incidence angle will be kept in a reasonable range at hub. The blockage caused by the recirculating flow reduces the suction entry diameter at the impeller inlet. Consequently, more head will be produced as soon as recirculation is fully developed. Therefore, recirculation is beneficial for a rising head curve at part load as discussed by Guelich [7], while Fraser et al. [8] have reported about excessive recirculation which has caused damage due to cavitation and pressure pulsation.

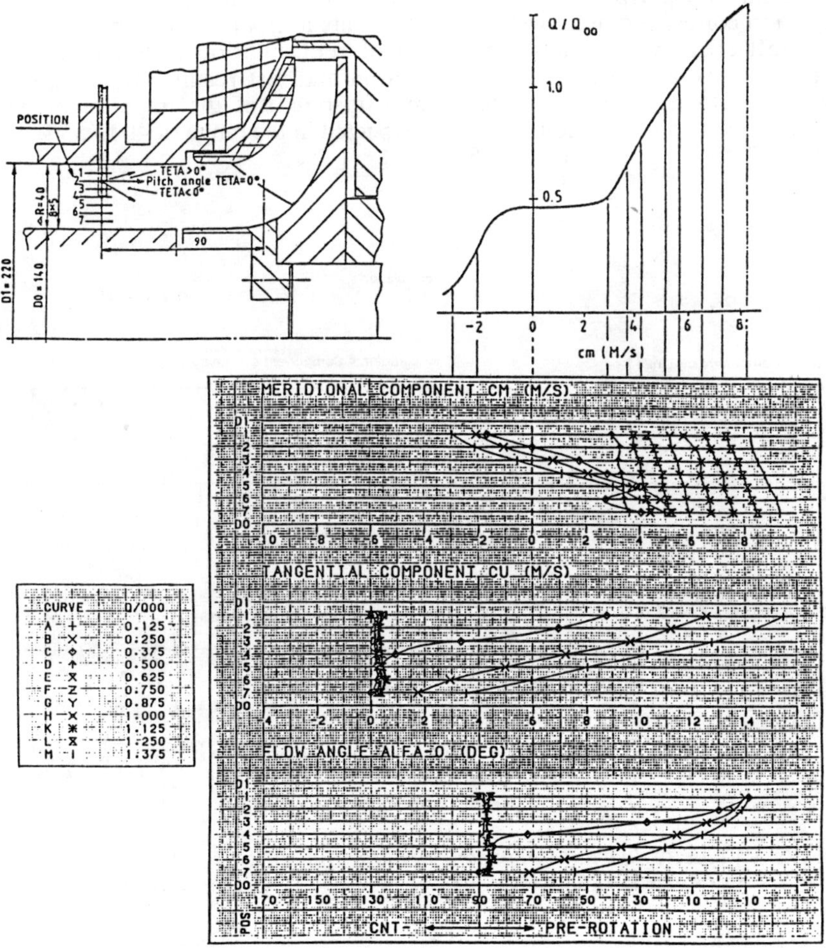

Fig. 6: Change of flow field in the suction line due to onset of impeller suction recirculation (published by Florjancic et al. [9])

Velocity profiles upstream of the impeller inlet

From overload to the onset of recirculation, the flow pattern remains nearly the same at the impeller inlet. It is defined mainly by the geometry and boundary conditions upstream of the impeller.

At the onset of recirculation an abrupt change of the flow field can be observed in the impeller eye. The fluid still enters the impeller at hub, but partly returns into the suction pipe at the impeller shroud (fig. 6). It was found that the extension of the recirculation zone is larger when viscous effects are present than for an ideal fluid.

At fully developed recirculation, the velocity profiles for both the meridional and peripheral component derived from the 3-D Euler flow calculation correspond quite well with test results in front of the impeller inlet (fig.7 left). But with smaller amounts of recirculation a considerable discrepancy exists between flow calculation and test results (fig.7 right). From the velocity measurements 90 mm upstream of the impeller blade leading edge the onset of recirculation is expected in the range of 40 to 55% of the best efficiency point flow (fig. 6).

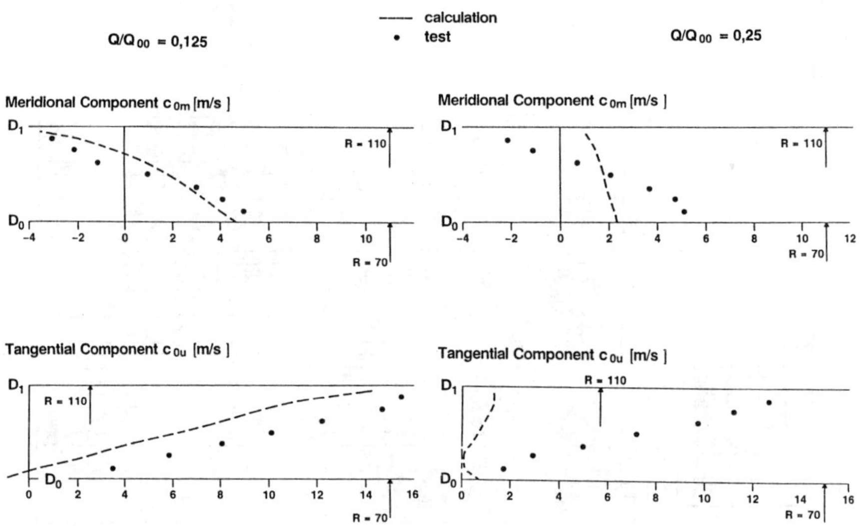

Fig. 7: Velocity distribution upstream of the impeller $Q/Qoo = 0.125$ (left), $Q/Qoo = 0.25$ (right)

At 25% load the calculated velocity profile in the impeller still shows a distribution similar to that at best efficiency, while the measurements indicate clearly the flow returning from the impeller shroud into the suction pipe (fig.7 right). The discrepancy is caused by the neglected leakage flow which has a strong peripheral component, and the boundary layer effects which are important at the onset of recirculation as well as by the uniform velocity with a zero peripheral component described at the inlet. Fig.9 shows clearly that recirculation exists also in the case of an inviscid fluid at 25% load. But, there is no backflow at a distance of 90 mm upstream of the impeller due to the smaller extension of the recirculation zone.

Velocity distribution within the impeller

Typical flow patterns for hub and shroud at the best efficiency point are shown in fig.2, where the flow follows the vaned channel quite well and where secondary flow effects are of minor importance.

The velocity vectors for fully developed recirculation have been plotted in fig.8. The flow enters the impeller at the hub and returns from the shroud into the suction pipe where it is forced to reenter the impeller. Only a fraction of the fluid which enters the impeller will be discharged into the diffuser from the vane pressure side.

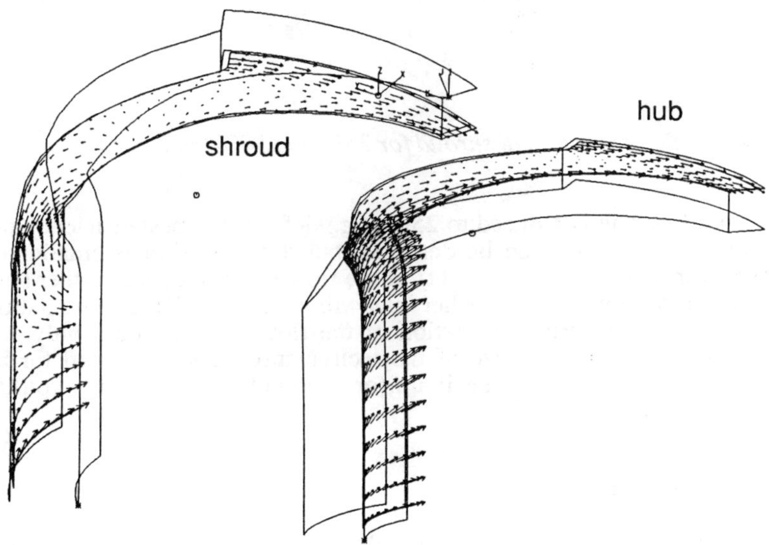

Fig. 8: *Velocity vectors for fully developed recirculation (12.5% load)*

A comparison of the flow patterns at best efficiency point (fig.2) and at fully developed recirculation (fig.8) shows clearly that the shroud suffers a lot more from recirculation than the hub. Therefore, only the velocity vectors at shroud will be considered for further flow rates.

At 50% flow rate a small dead water zone is formed at the vane suction side of the inlet region as shown in fig. 9 right, but there is no backflow which returns into the suction pipe.

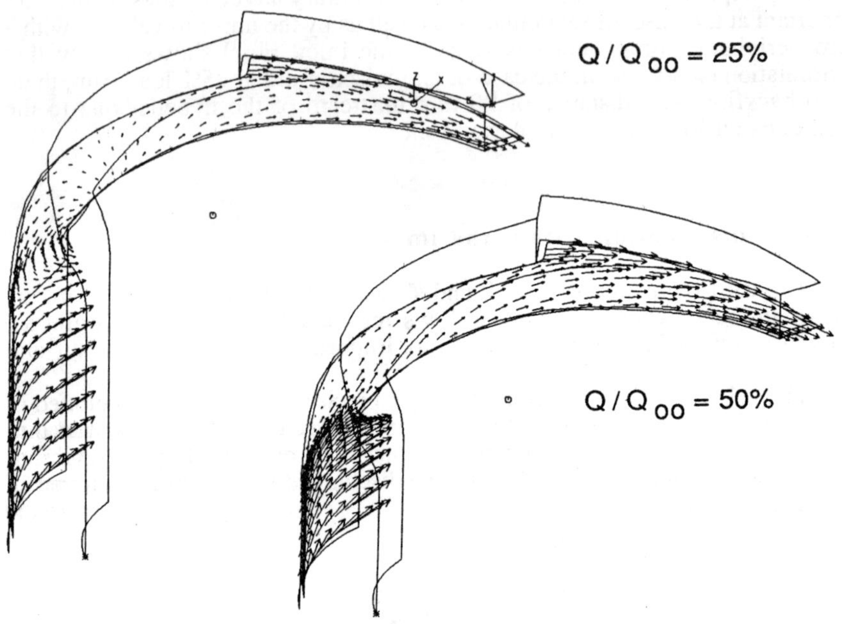

Fig. 9: Velocity vectors at shroud for 25% and 50% load

If the flow rate is reduced to 25% (fig.9 left) of the best efficiency point flow rate, recirculation can be detected, but the extension is smaller than observed during the test (fig.7 right). If the flow rate is reduced further (fig.8) the recirculation zone at the impeller inlet will increase until the recirculation is fully developed. A further reduction of the flow rate produces only small changes. It seems that the size of the recirculation zone is limited for both viscous and ideal fluids, even if uniform boundary conditions would be applied at infinity.

Part load mechanisms

The load at vane inlet depends roughly on the square of the radial leading edge position. This is the reason why recirculation starts even for inviscid flow at the shroud. It seems that the vane load distribution along the leading edge is the most important factor for part load flow recirculation. In the case of a viscous fluid losses, due to deceleration and due to the change of the flow direction, increase also with the axial distance. This might be one explanation why the recirculation starts at larger flow rates in the case of a real fluid than under ideal conditions.

A similar effect will be produced by the leakage flow in the lateral recess between shroud and casing from the impeller exit to the suction pipe, which increases the prerotation at the inlet. The description of a constant meridional velocity and a zero peripheral component at a large distance upstream of the impeller is probably an additional reason for the relative small extension in the case of the ideal fluid, because all velocity traverses for flow rates above the onset of backflow show a small peripheral component and a meridional velocity which decreases with the radial distance.

Due to the blockage effect and prerotation caused by the recirculation at the shroud, the incidence angle at hub will be improved. As soon as recirculation is fully developed the head will increase progressively against shut off due to the reduction of the outer impeller inlet diameter for the throughflow. At over load no such mechanisms exist, because the inlet conditions would even become worse if the inlet area would be blocked by recirculation.

Conclusions

The ideal head of an impeller can be predicted with high accuracy by means of a 3-D Euler-code. The total band width for 21 tests was 11.2%, while the best slip factor correlation method yields a fluctuation range of 18%. Therefore the fully 3-D Euler flow calculation is a major step to obtain more precise estimates during the design process, because it takes the whole geometry into account. It was found that the correct formulation of boundary conditions is very important for a high accuracy.

Part load flow calculations have been performed in order to check the limitations of the program. It was surprising that the time marching method worked even at zero load and produced reasonable results with velocity distributions comparable to flow measurements and flow observations.

Quantitative comparisons at flow rates with fully developed recirculation have shown a good agreement with measured velocity profiles upstream of the impeller (fig. 7 left). Only at onset of recirculation (fig. 7 right) a discrepancy exists, which might be explained by the leakage flow and boundary layer effects, because the program does not account for these phenomena.

Literature

1. *E. Goede, I.L. Ryhming*
3-D Computation of the Flow in a Francis Runner
Sulzer Technical Review 4, vol. 68, 1987

2. *F. Thibaud, A. Drotz, G. Sottas*
Validation of an EULER code for hydraulic turbines
AGARD Symposium, 1988, Lisbon

3. *E. Goede, R. Cuénod, R. Grunder, J. Pestalozzi*
A New Computer Method to Optimize Turbine Design and Runner Replacement. Hydro Review, 1991

4. *E. Goede, R. Cuénod, H. Keck, J. Pestalozzi*
3-Dimensional Flow Simulation in a Pump-Turbine, ASME, FED-Vol 86, Industrial and Agricultural Applications of Fluid Mechanics, Book No.H0055-1989

5. *A.J. Chorin*
A Numerical Method for Solving Incompressible Flow Problems
J.Comp. Phys. 2, 1967, pp.12-26

6. *R. Peyret, T. Taylor*
Computational Methods for Fluid Flow
Springer, New York, 1982

7. *J. Guelich*
Remarks on Stability of Characteristic Curves of Centrifugal Pumps
Pumpentagung Karsruhe 1986

8. *W.H. Fraser, I.J. Karassik, A.R. Bush*
Study of Pump Pulsation, Surge and Vibration Throws Light on Reliability vs Efficiency. Power, August 1977

9. *D. Florjancic, V. Bolleter, A. Simon*
Boiler Feed Pumps for Modern Fossil
Fired Power Plants, Sulzer, 1987

SECTION 2: AERODYNAMICS

Experience with Transonic Flow IE Computations

H. Hu

Department of Mathematics, Hampton University, Hampton, Virginia 23668, U.S.A.

ABSTRACT

An integral equation method based on the full-potential equation for transonic flow calculations is presented. The full-potential equation is written in the moving frame of reference, in the form of the Poisson's equation. The integral equation solution in terms of the velocity field is obtained by the Green's theorem. The mixed nature of the transonic flow is treated by a type- difference scheme. The numerical solutions are obtained by a time-marching (if unsteady flows), iterative procedure. The computational examples presented in the present paper include steady and unsteady, two-dimensional (airfoil) and three-dimensional (wing) flows. The method of combining the integral equation solution with the finite-volume Euler solution is also presented. Through studying the method and their computational examples, the capabilities and limitations of the transonic integral equation method are discussed. Finally, the needs for further research is addressed.

Key Words: integral equation, field/boundary elements, full-potential equation, transonic flow.

INTRODUCTION

Starting in 1970, a great deal of progress has been made in solving transonic flow by using the finite-difference method (FDM) and finite-volume method (FVM). Although the FDM and FVM are successful in dealing with transonic flows, there are several drawbacks associated with those methods. In the FDM and FVM, fine grid points are needed over a large computational domain. Moreover, there are major technical difficulties in generating suitable grids for complex three-dimensional aerodynamic configurations.

On the other hand, the integral equation method (IEM, or called Field- Boundary Element Method, Field-Panel Method) has several advantages over the FDM and FVM. The IEM involues evaluation of integrals, which is more accurate and simpler than the FDM and FVM. The IEM automatically satisfies the far-field boundary conditions and hence only a small limited computational domain is needed. The IEM does not suffer from the artificial viscosity effects as compared to FDM and FVM for shock capturing in transonic flow computations. Moreover, the generation of the three-dimensional grid (field-elements) for complex configuration is not difficulty in the IEM, since the surface fitted grid is not required.

Because of these advantages of IEM, it is highly desirable to fully develop the IEM to treat transonic flows. Integral equation methods for transonic flows have been developed by several investigators[1–8]. The author and his co-workers have been devoted to the development of the IEM for steady and unsteady transonic airfoil and wing flow computations during the past several years[9–15]. In the present paper, the recent development along with the computational examples are presented. Through studying the method and the numerous computational examples, the capabilities and limitations of the transonic integral equation method are discussed. Finally, the needs for further research are addressed.

FORMULATION

Full-Potential Equation

In the space-fixed frame of reference, the continuity and momentum equations for unsteady, inviscid compressible flows with negligible body forces are given by

$$\frac{D\rho}{Dt} + \rho\nabla\cdot\vec{V} = 0 \tag{1}$$

and

$$\rho\frac{D\vec{V}}{Dt} + \nabla p = 0 \tag{2}$$

respectively, where ρ is the density; p the pressure; $\vec{V}$ the absolute velocity; and t the time.

However for a general unsteady motion of a body, the governing equations are simple to solve if the moving (body-fixed) frame of reference formulation is used. This formulation does not require the grid-motion calculation since the grid is rigidly fixed in the frame of reference. In addition to the space-fixed frame of reference $OXYZ$, a moving frame of reference $oxyz$ is introduced as shown in Fig. 1. The moving frame of reference $oxyz$ is translating at a velocity of $\vec{V}_o(t)$ and rotating around a pivot point, $\vec{r}_p = (x_p, y_p, z_p)$, at an angular velocity of $\vec{\Omega}(t)$.

The relation for the absolute velocity, relative velocity ($\vec{V}_r$) and transformation velocity $\vec{V}_e = (\vec{V}_o + \vec{\Omega} \times (\vec{r} - \vec{r}_p))$ is given by

$$\vec{V} = \vec{V}_r + \vec{V}_e \tag{3}$$

where $\vec{r}$ is the position vector measured in the moving frame of reference.

The substantial derivative of a scalar quantity like ρ is related to its substantial derivative in the moving frame and to the local derivative in the moving frame by the equation

$$\frac{D\rho}{Dt} = \frac{D'\rho}{Dt} = \frac{\partial'\rho}{\partial t} + \vec{V}_r \cdot \nabla\rho \tag{4}$$

where the ($\prime$) refers to the derivative with respect to the moving frame of reference. Also, the substantial derivative of a vector quantity like $\vec{V}$ is realted to one in moving frame of reference by

$$\frac{D\vec{V}}{Dt} = \frac{D'\vec{V}}{Dt} + \vec{\Omega} \times \vec{V} = \frac{\partial'\vec{V}}{\partial t} + \vec{V}_r \cdot \nabla\vec{V} + \vec{\Omega} \times \vec{V} \tag{5}$$

By using Eqs. (3)-(5) and assuming that the flow is isentropic and irrotational, Equations (1) and (2) take the form in the moving frame of reference as follows:

$$\nabla^2\Phi = -\frac{\nabla\rho}{\rho} \cdot \vec{V}_r - \frac{1}{\rho}\frac{\partial'\rho}{\partial t} \tag{6}$$

and

$$\frac{\rho}{\rho_\infty} = \left\{1 - \frac{\kappa - 1}{2a_\infty^2}[(\nabla\Phi)^2 + 2(\frac{\partial'\Phi}{\partial t}) - 2\nabla\Phi \cdot \vec{V}_e]\right\}^{\frac{1}{\kappa-1}} \tag{7}$$

where Φ is the absolute velocity potential given by $\vec{V} = \nabla\Phi = \nabla'\Phi$ and κ is the gas specific heat ratio. Equation (6) is the unsteady full-potential equation in the moving frame of reference with the density given by Eq. (7). After introducing the characteristic parameters of length (such as, airfoil chord length), speed of sound at infinity (a_∞) and density at infinity (ρ_∞) and defining the translation mach number as $M_o = \mid \vec{V}_o \mid / a_\infty$ (note that $M_o = M_\infty$, free-stream Mach number), Eqs. (6) and (7) takes dimensionless form as follows:

$$\nabla^2\Phi = G \tag{8}$$

with

$$G = G_1 + G_2 \tag{9}$$

$$G_1 = -\frac{\nabla\rho}{\rho} \cdot \vec{V}_r \tag{10}$$

$$G_2 = -\frac{1}{\rho}\frac{\partial'\rho}{\partial t} \tag{11}$$

and

$$\rho = \left\{1 + \frac{\kappa - 1}{2}[-V_r^2 + (\vec{V}_o + \vec{V}_e)^2 - 2(\frac{\partial'\Phi}{\partial t})]\right\}^{\frac{1}{\kappa-1}} \tag{12}$$

where G_1 is the compressibility and G_2 is the unsteadiness. It should be noticed that all the quantities in Eqs. (8)-(12) are dimensionless, although the same notation as ones used for dimensional quantities are used.

For steady flows, $\vec{\Omega}(t)$, G_2 and $\partial'\Phi/\partial t$ are set to be zero; and $\vec{V}_o(t)$ becomes time-independent, which is the negative of the free-stream velocity.

Boundary Conditions

The boundary conditions are surface no-penetration condition, Kutta condition, infinity condition, wake kinematic condition and wake dynamic condition, which are described as follows:

$$\vec{V}_r \cdot \vec{n}_g = 0 \quad \text{on} \quad g(\vec{r}) = 0 \tag{13}$$

$$\Delta C_p \mid_{sp} = 0 \tag{14}$$

$$\nabla\Phi \to 0 \quad \text{away from} \quad g(\vec{r}) = 0 \quad \text{and} \quad w(\vec{r}, t) = 0 \tag{15}$$

$$\frac{1}{\mid \nabla w \mid}\frac{\partial' w}{\partial t} + \vec{V}_r \cdot \vec{n}_w = 0 \quad \text{on} \quad w(\vec{r}, t) = 0 \tag{16}$$

and

$$\Delta C_p = 0 \quad \text{on} \quad w(\vec{r}, t) = 0 \tag{17}$$

where $\vec{n}$ is the surface unit normal vector; the subscripts g and w refer to the body (wing or airfoil) and wake surface of $g(\vec{r}) = 0$ and $w(\vec{r}, t) = 0$, respectively; ΔC_p is the pressure jump across the surface; and the subscript sp refers to edge of separation.

Integral Equation Solution

By using the Green's theorem, the integral equation solution of Eq. (8) in terms of the relative velocity field is given by

$$\begin{aligned} \vec{V}_r(x, y, z, t) = & -\vec{V}_o(t) - \vec{\Omega}(t) \times (\vec{r} - \vec{r}_p) \\ & - \frac{1}{4\pi}\int\int_g \frac{q_g(\xi, \eta, \zeta, t)}{d^2}\vec{e}_d ds \\ & + \frac{1}{4\pi}\int\int_g \frac{\vec{\gamma}_g(\xi, \eta, \zeta, t) \times \vec{d}}{d^3} ds \\ & + \frac{1}{4\pi}\sum_{iw=1}^{NW}\int\int_w \frac{\vec{\gamma}_w(\xi, \eta, \zeta, t) \times \vec{d}}{d^3} ds \\ & + \frac{1}{4\pi}\int\int\int_V \frac{G(\xi, \eta, \zeta, t)}{d^2}\vec{e}_d d\xi d\eta d\zeta \\ & + \frac{1}{4\pi}\int\int_S \frac{q_S(\xi, \eta, \zeta, t)}{d^2}\vec{e}_d ds \end{aligned} \tag{18}$$

where q is the surface source distribution; $\vec{\gamma}$ is the surface vorticity distribution; the subscript S refers to the shock surface; the index NW is the total number of wake surfaces; ds is the infinitesimal surface area; the vector $\vec{d}$ is given by $\vec{d} = (x - \xi)\vec{i} + (y - \eta)\vec{j} + (z - \zeta)\vec{k}$; and $\vec{e}_d$ is defined by $\vec{e}_d = \vec{d}/|\vec{d}|$.

It should also be noticed that Eq. (18) has been written for three-dimensional flows. For two-dimensional flows, above surface integrals become line integrals and the volume integrals become field surface integrals. The coordinates, z and ζ, are not used. The coefficients of $1/4\pi$ are replaced by $1/2\pi$ for two-dimensional flows. The last integral term in Eq. (18) is used only for the shock-fitting solutions in steady two-dimensional flows.

COMPUTATIONAL SCHEME

A sketch of the IE computational domains is shown in Fig.2 for three-dimensional flows. Due to the nature of the nonlinearity of the flow, the solutions are obtained through a iterative procedure where the compressibility (G_1), unsteadiness (G_2), and the wake shape and its strength are updated through each iteration. Once the solution converges, it marches to the next time step. The details of the solution procedure can be found in Ref. [9,14]. Here only the treatment of the mixed nature of the flow and its shock is described.

Type-Difference Scheme

Transonic flows are characterized by the presence of both subsonic and supersonic regions within the flow field simultaneously. Therefore, transonic flows are described by a mixed elliptic-hyperbolic partial differential equation with the boundary between them unknown apriori. To be consistent with the mixed nature of the transonic flow, the Murman-Cole type-difference scheme is applied to the present IE computation for $\nabla\rho$ calculations. For the subsonic points where the local Mach number is less than unit, central- differencing is used. For supersonic points where the local Mach number is greater than unit, backward-(upstream-) differencing is used. One exception is to use forward-differencing at the first elements after the shock discontinuity. This type-difference scheme is consistent with the nature of the transonic flow, because the local disturbance in a subsonic flow propagates in all directions while in a supersonic flow the local disturbance is confined to the downstream Mach cone of the disturbance.

Shcok-Fitting Technique

It should be mentioned that mathematically the fourth volume integral term of Eq. (18) includes all compressibility effects including shock discontinuity. Since a relative coarse grid has been used in the present IE computational domain, the contribution of the shock discontinuity is extracted from the fourth integral terms and it is represented explicitly by fifth integral term of Eq. (18). The strength of the shock panel (q_S) is equal to the difference of normal velocity across the shock. The slope of the shock panel is determined by the Rankine-Hugoniot relation. The technique to introduce the shock panel term, the fifth integral term of Eq. (18), and to fit the shock using Rankine-Hugoniot relation is called shock-fitting technique, as shown in Fig. 3. This shock-fitting technique is applied to steady 2-D computations.

NUMERICAL EXAMPLES

The integral equation method has been applied to steady and unsteady, two- dimensional and three-dimensional transonic flows. The computational results along with the experimental data and other computational results are presented in the following sub-sections.

Steady Subsonic Airfoil Flow

The computational results for a steady compressible shock-free flow at high subsonic Mach number are presented here as the first numerical example. The purpose is to validate the IEM for nonlinear compressible flows. The surface pressure distribution[9,10] is shown in Figure 4, along with the finite-difference (FD) Euler solution[16]. In this case, the NACA0012 airfoil has been used at the flow condition of $M_\infty = 0.72$ and $\alpha = 0^o$. The results show that the integral equation (IE) solutions for this case agree with the FD Euler solutions.

Steady Transonic Airfoil Flow

The second numerical example is a transonic flow case with shock of moderate strength[9,10]. Figure 5 shows the results for the flow around the NACA64A010A airfoil at $M_\infty = 0.796$ and $\alpha = 0^o$. This airfoil has 10% thickness and a small camber. In this case, a computational domain of 2×1.5 airfoil chord length with 64×60 field-elements has been used. The number of iteration used to achieve convergence is about 25. This is much less than that used in FDM and FVM. The results show that the integral equation method can predict the shock correctly.

Unsteady Transonic Airfoil Flow

The third numerical example is the unsteady transonic airfoil flow case. In this case, the unsteady IEM has been applied to the NACA0012 airfoil at a translation Mach number of 0.755 undergoing forced pitching oscillation around a pivot point at the quarter-chord, measured from the leading edge ($\vec{r_p} = (0.25, 0.0)$). The angles of attack are given by: $\alpha(t) = 0.016^o + 1.255^o sin(0.1632t)$. Figure 6 shows the computed surface lifting coefficients[9,11] along with the finite-volume (FV) Euler solutions produced by Ref. [18]. The comparison shows a good agreement between IE solutions and FV Euler solutions. The shock strength and its motion are also predicted correctly (see Ref. [9,11]).

Unsteady Transonic Wing Flow

Recently, IEM has been applied to three-dimensional unsteady transonic flows. The unsteady transonic flow around a rectangular wing has been computed. In this numerical example, an zero-thickness rectangular wing with aspect ratio of 2 is given an forced transient pitching motion at a transonic translation Mach number of 0.7. To simplify the problem, only a pitching motion in xy-plane is considered; and therefore $\vec{\Omega}$ is given by $\vec{\Omega} = 0\vec{i} + 0\vec{j} + \alpha\vec{k}$, where α is given by $\alpha(t) = 5^o + 0.5^o t$ with $t = n\Delta t$. n is time-step index (from 1 to 10 in this case) and Δt is time-step size, which is chosen as 1. The half-span of the wing and wake is divided into 10×6 and 10×10 quadrilateral panels, respectively. Each quadrilateral panel consists of two triangle panels. The time history of the surface pressure distributions over the wing root section is shown in Figure 7. The results show that the IEM can capture the unsteady effect correctly, although finer field-elements may be required to narrow the shock region and to predict the shock motion as accurate as possible.

CAPABILITIES AND LIMITATIONS

The steady and unsteady integral equation methods for nonlinear compressible flows have been developed. The methods have been applied to steady airfoil, unsteady airfoil and unsteady wing flows with or without shocks. The comparison of the present solutions with experimental data and FD or FV solutions shows that the integral equation methods based on the linear theorem can handle nonlinear flow problems accurately. For transonic flows with shocks of weak to moderate strength, IEM predicts shocks correctly, with the exception of slight underprediction of the shock strength (Fig. 5). For unsteady flows, the motion of the shock agrees with that predicted by FV Euler solutions (see Ref. [9,11]), and the predicted lifting ceofficient agrees with one obtained by FV Euler computation (Fig. 6).

The advantages of the small computational domain and coarse grid have been utilized in the present IEM. For airfoil flow computations, a computational domain of 2×1.5 airfoil chord length with 64×60 field-elements has been used. The use of less number of field-elements with larger field- elements over the outer region inside the domain is possible. For wing flow computations, a computational domain of $2.3 \times 0.75 \times 1.5$ wing root chord length with $23 \times 9 \times 9$ field-elements has been used, although finer field-elements around the shock region may be required to accurately predict the shock location and its strength.

The numbers of iteration used for steady flow computations are about 25 for airfoil and 5 for wing flows, respectively. This is much less than those used in FD and FV computations (usually order of 10^3). For unsteady flows, the numbers of iteration used in each time step range from 1 to 3. Large time steps have also been used in the present unsteady flow computations. This is also the one of the advantages of IEM. For a whole cycle of pitching oscillation for example, a total of 36 time steps has been used; while a typical implicit FD or FV computation needs about 500 time steps for the same case. Therefore, IEM is nevertheless efficient in terms of the number of iteration and the time step size, as compared to existing FDM and FVM.

By examining the numerical examples presented here it is found that, on the other hand, all these computations are restricted to the flows with shocks of weak to moderate strength. As the best of the author's knowlendge, all existing integral equation solutions based on potential flow formulation for transonic flows are restricted to flows without strong shocks. The potential flow assumption neglects the effects due to viscosity, vorticity and entropy production. For transonic flows with strong shocks and masive separation, the potential flow assumption is not an adequate approximation to the real flow.

In order to extend the integral equation method to a wide range of transonic flows, the effects of viscosity, vorticity and entropy production must be considered. The attempt has been made by some inveatigators by using the Navier-Stokes equations in integral equation formulation. On the other hand, the idea of combining FVM for Euler equations (neglecting viscous effect) with IEM for full-potential equation has been developed[9,10]. In the following section, a brief description of the method with the computational examples will be given.

HYBRID IE-FV METHODS

A hybrid IE-FV method has been developed for both steady and unsteady transonic airfoil flow computations. In this method, unsteady Euler equations are solved in the small inner domain using a FVM, while the full-potential equation is solved over outer domain using IEM to update the outer boundary conditions of the Euler inner domain. A sketch of the computational domain for unsteady flow computation is shown in Fig.8. The method has been applied to both steady and unsteady transonic airfoil flows.

Figure 9 shows the computational domain and numerical results[9,10] for a typical strong shock flow case, the steady flow around NACA0012 airfoil at $M_\infty = 0.84$ and $\alpha(t) = 0^o$. For steady flows as shown in Fig.9, the Euler computational domain is smaller than one for unsteady flows, since shock is fixed in location for steady flows. This case took 10 IE-iterations to locate the shock, 300 FV-iterations to achieve a residual error of 10^{-3} and 3 IE-iterations to update the Euler inner domain outer boundary conditions. The present solution compares very well with the FV Euler solution[19] both in strength and location of the shock (a strong shock !).

Figure 10 shows the recently computed lifting coefficients for a transient ramp motion of the NACA0012 airfoil at $M_o = 0.56$ and angles of attack given by: $\alpha(t) = -0.01^o + 0.855^o t$, along with the experimental data of Ref. [20]. The comparison shows a good agreement between the present solution and the experimental data at a range of low to moderate angles of attack (from 0° to 4.7^o). As the angle of attack increases, the discrepancy between the present solution and the experimental data increases. This difference may be attributed to the absence of the viscous terms including the turbulence effects from the present inviscid formulation - Euler/potential formulation.

CONCLUDING REMARKS

The integral equation solutions of the full-potential equation for transonic flows have been developed. The numerical examples have been presented. Throught studying the methods and their numerical examples, the capabilities and limitations of the transonic integral equation methods based on full- potential formulation have been discussed. The combination of full-potential IE and Euler FV solutions has benn demonstrated for a strong shock flow case and an unsteady transonic flow case. Further works of extending the present transonic integral equation method may be focused on: (1) whole aircraft computations; (2) three-dimensional, general unsteady motion (including deformation) computations; (3) two-dimensional and three-dimensional, steady and unsteady flow computations by hybrid FV

Navier-Stokes/IE potential method.

ACKNOWLEDGEMENT

This research work has been partially supported by NASA Langley Research Center under the Grant No. NAG-1-1170. Dr. E. Carson Yates, Jr. is the technical monitor.

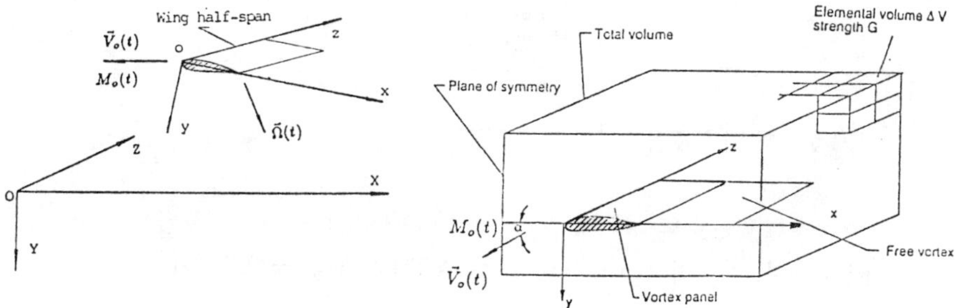

Figure 1. Frames of reference. Figure 2. IE Computational domain.

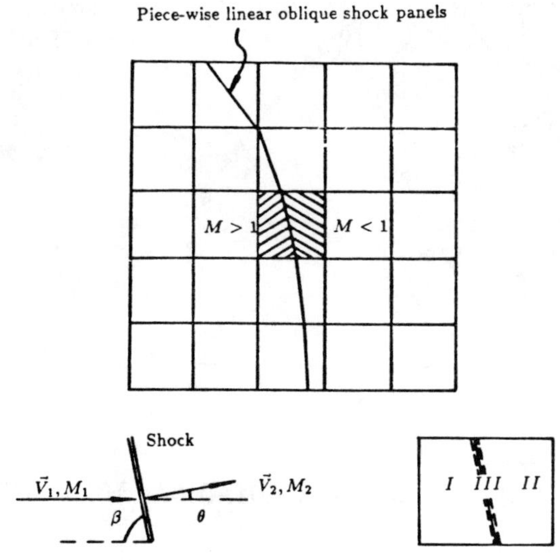

Figure 3. Illustration of shock panels and field-element splitting; Areas I and II: 4th integral term; Area III: 5th integral term of Eq. (18).

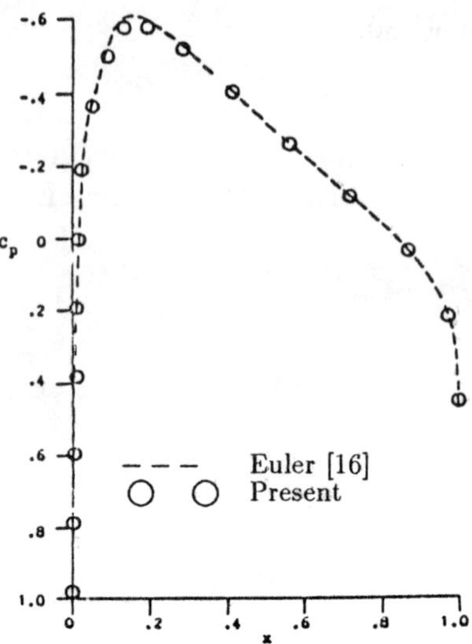

Figure 4. C_p distribution, NACA0012, $M_\infty = 0.72$, $\alpha = 0^o$.

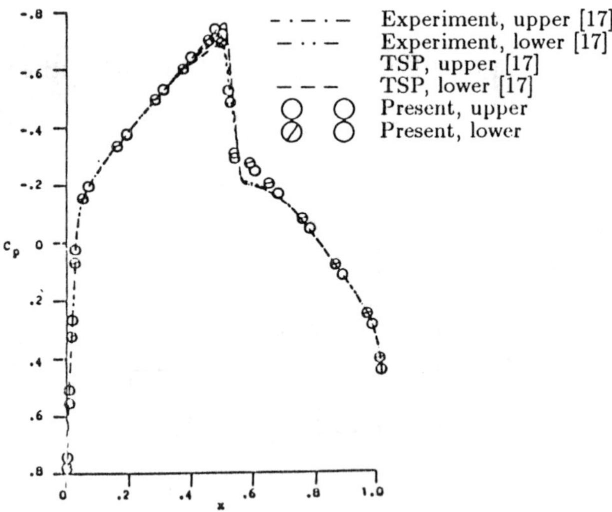

Figure 5. C_p distribution, NACA64A010A, $M_\infty = 0.796$, $\alpha = 0^o$.

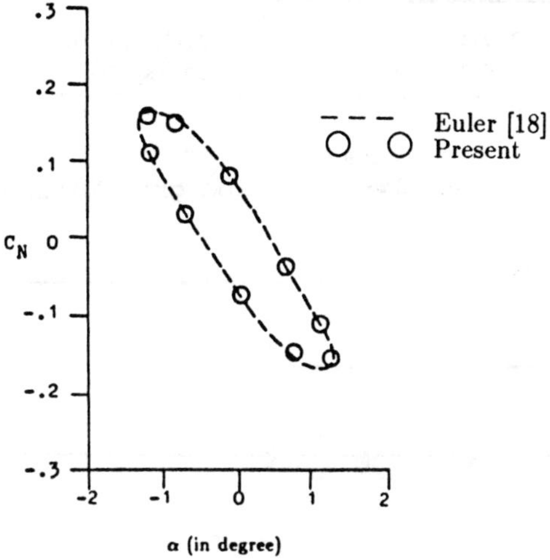

Figure 6. Lifting coefficients, pitching oscillation, $M_o = 0.755$, $\alpha(t) = 0.016^o + 1.255^o sin(0.1632t)$.

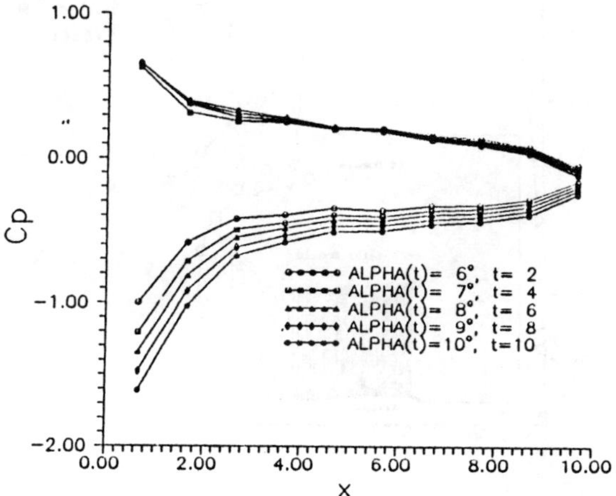

Figure 7. Wing root section C_p history, ramp motion, $M_o = 0.7$, $\alpha(t) = 5^o + 0.5^o t$.

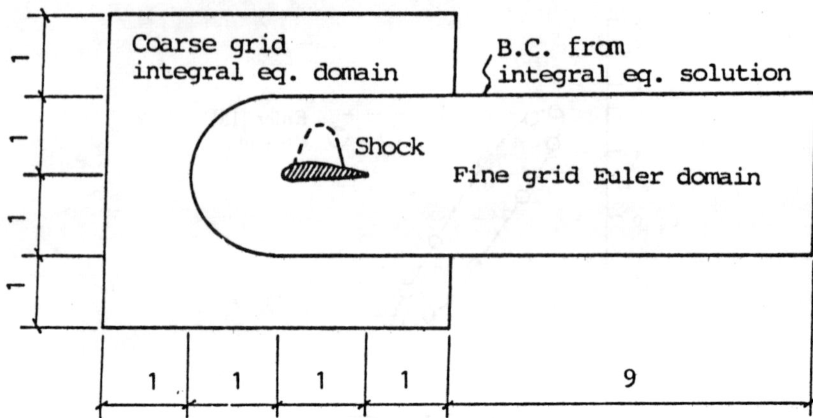

Figure 8. Hybrid IE-FV computational domain for unsteady flows.

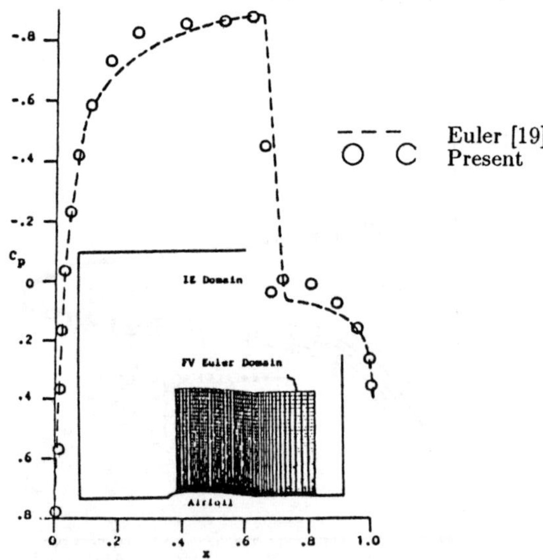

Figure 9. Euler domain and C_p distribution, a strong shock case, steady flow, $M_\infty = 0.84$, $\alpha = 0$.

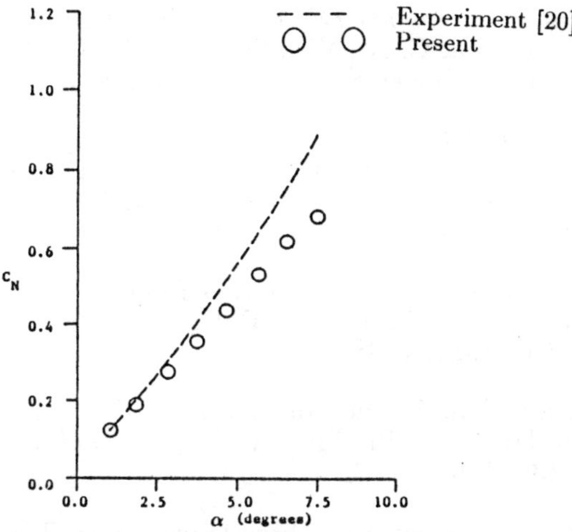

Figure 10. Lifting coefficients, ramp motion, $M_o = 0.56$, $\alpha(t) = -0.01^o + 0.855^o t$.

REFERENCES

1 Piers, W. J. and Sloof, J. W. Calculation of Transonic Flow by Means of a Shock-Capturing Field Panel Method, AIAA Paper 79-1459, 1979.

2 Hounjet, M. H. L. Transonic Panel Method to Determine Loads on Oscillating Airfoils with Shocks, *AIAA Journal*, Vol. 19, No. 5, pp.559-566, 1981.

3 Tseng, K. and Morino, L. Nonlinear Green's Function Methods for Unsteady Transonic Flows, in *Transonic Aerodynamics*, (Ed. Nixon, D), AIAA, pp.565-603, 1982.

4 Oskam, B. Transonic Panel Method for the Full Potential Equation Applied to Multicomponent Airfoils, *AIAA Journal*, Vol. 23, No. 9, pp.1327-1334, 1985.

5 Erickson, L. L. and Strande, S. M. A Theoretical Basis for Extending Surface-Paneling Methods To Transonic Flow, *AIAA Journal*, Vol.23, No. 12, pp.1860-1867, 1985.

6 Sinclair, P. M. An Exact Integral (Field Panel) Method for the Calculation of Two-Dimensional Transonic Potential Flow Around Complex Configurations, *Aeronautical Journal*, Vol. 90, No. 896, pp.227-236, 1986.

7 Kandil, O. A. and Yates, E. C., Jr. Computation of Transonic Vortex Flows Past Delta Wings-Integral Equation Approach, *AIAA Journal*, Vol. 24, No. 11, pp.1729-1736, 1986.

8 Ogana, W. Boundary Element Methods in Two-Dimensional Transonic

Flows, in *Boundary Elements IX, Vol.3: Fluid Flow and Potential Applications*, (Eds. Brebbia, C. A., Wendland, W. L. and Kuhu, G.), Computational Mechanics Publications, Springer-Verlag, pp.567-582, 1987.

9 Hu, H. Full-Potential Integral Solutions for Steady and Unsteady Transonic Airfoils with and without Embedded Euler Domains, Ph.D. Dissertation, Dept. of Mechanical Engineering and Mechanics,Old Dominion University, Norfolk, Virginia, March 1988.

10 Kandil, O. A. and Hu, H. Full-Potential Integral Solution for Transonic Flows with and without Embedded Euler Domains, *AIAA Journal*, Vol. 26, No. 9, pp.1074-1086, 1988.

11 Kandil, O. A. and Hu, H. Integral Solution of Unsteady Full- Potential Equation for a Transonic Pitching Airfoil, *Journal of Aircraft*, Vol. 27, No. 2, pp.123-130, 1990.

12 Hu, H. and Chu, L-C. Unsteady Three-Dimensional Transonic Flow Computations Using Field Element Method, in *Boundary Element Methods in Engineering*, (Eds. Annigeri, B. S. and Tseng, K.), Springer-Verlag, pp. 140-146, 1990.

13 Hu, H. and Kandil, O. A. Unsteady Field Element Method with Embedded Euler Domain for Transonic Pitching Airfoil Flows, in *Boundary Element Methods in Engineering*, (Eds. Annigeri, B. S. and Tseng, K.), Springer-Verlag, pp. 161-168, 1990.

14 Hu, H. Unsteady Transonic Wing Flow Computations Using Field- Boundary Element Methods, to appear in *Journal, Engineering Analysis with Boundary Elements.*

15 Hu, H. Application of Integral Equation Method to Flows Around a Wing with Circular-Arc Section, in *Boundary Elements XIII*, (Eds. Brebbia, C. A. and Gipson, G. S.), Computational Mechanics Publications, Elsevier Applied Science, pp. 209-217, 1991.

16 Sells, C. L. Plane Subcritical Flow Past a Lifting Airfoil, in *Proceedings of the Royal Society of London*, Series A, Vol. 308, No. 1494, pp.377-401, 1969.

17 Edwards, J. W., Bland, S. R. and Seidel, D. A. Experience with Transonic Unsteady Aerodynamic Calculations, NASA TM-86278, 1984.

18 Kandil, O. A. and Chuang, H. A. Unsteady Transonic Airfoil Computation Using Implicit Euler Scheme on Body-Fixed Grid, *AIAA Journal*, Vol. 27, No. 8, pp. 1031-1037, 1989.

19 Jameson, A., Schmidt, W. and Turkel, E. Numerical Solutions of the Euler Equations by Finite-Volume Methods Using Runge Kutta Time-Steping Scheme, AIAA Paper 81-1259, 1981.

20 London, R. H. NACA0012 Oscillatory and Transient Pitching, AGARD R- 702, pp.3.1-3.25, 1982.

A 3D IEM for Compressible Wing Flows With and Without Shocks

H. Hu

Department of Mathematics, Hampton University, Hampton, Virginia 23668, U.S.A.

ABSTRACT

An integral equation (or called field-panel, field-boundary element) scheme for solving the full-potential equation for incompressible and compressible flows with and without shocks has been developed. The full-potential equation has been written in the form of the Poisson's equation. Compressibility has been treated as non-homogeneity. The integral equation solution in terms of velocity field is obtained by the Green's theorem. The solution consists of wing (or a general body) surface (boundary elements) integral term(s) of vorticity/source distribution(s), wake surface (boundary elements) integral term(s) of free-vortex sheet(s), a volume (field-elements) integral term of compressibility over a small limited domain around the source of disturbance, and a shock surface (boundary elements) integral term of source distributions. Solution is obtained through an iterative procedure for non-linear compressible flows. To be consistent with the mixed-nature of transonic flows, the Murman-Cole type-difference scheme is used to compute the derivatives of the density. The present scheme is applied to flows around a rectangular wing with circular-arc section at incompressible, high- subsonic and transonic flow conditions.

Key Words: integral equation method, full-potential equation, subsonic and transonic wing flows.

INTRODUCTION

The finite-difference method (FDM) and finite-volume method (FVM) for solving transonic flows have been well developed during the past twenty years. Although the Navier-Stokes equation formulation for the transonic flow computations has been understood as the best model and the FDM and FVM are successful in dealing with transonic flows, the computation of the unsteady Navier-Stokes equations over complex three-dimensional configurations is very expensive, particularly for time-accurated unsteady flow computations. There are also major technical difficulties in FDM and FVM for generating suitable grids for complex three-dimensional aerodynamic configurations.

The experience has shown that rather accurate solutions can been obtained for many transonic flows using the inviscid modeling of the full-potential equation. For transonic flows without strong shocks and massive separations, the full-potential equation is an adequate approximation to the Navier-Stokes equations. The integral equation method (IEM) for the potential equation is an alternative to the FDM and FVM. Moreover, the IEM has several advantages over the FDM and FVM. The IEM involues evaluation of integrals, which is more accurate and simpler than the FDM and FVM, and hence a coarse grid (field-elements) can been used in IEM. The IEM automatically satisfies the far-field boundary conditions and therefore only a small limited computational domain is needed. The IEM does not suffer from the artificial viscosity effects as compared to FDM and FVM for shock capturing in transonic flow computations. The generation of the three-dimensional grid for complex configuration is not difficulty in the IEM, since the mapping from physical plane to computational plane is not required.

Integral equation methods for transonic flows have been developed by several investigators[1−18] during the past few years for steady airfoil, wing and aircraft configurations and unsteady airfoils and wings. In the present paper a method for computing general steady 3-D flows, which is an extension of 2D method of Ref.10, is presented along with simple numerical examples to demonstrate the capability and the potential of the present IE scheme for incompressible, subsonic and transonic flow computations. The shock-fitting technique is applied to the present transonic flow calculations, so that a coarse grid can be used.

FORMULATION

Governing Equations

The non-dimensional steady full-potential equation is given by:

$$\nabla^2\Phi = G \tag{1}$$

with

$$G = -\frac{\nabla\rho}{\rho}\cdot\vec{V} \tag{2}$$

and

$$\rho = [1 + \frac{\kappa - 1}{2}(1- \mid \vec{V} \mid^2)]^{\frac{1}{\kappa-1}} \tag{3}$$

where the characteristic parameters, ρ_∞, a_∞ and l have been used; a is the speed of the sound, ρ the density, and l the wing surface panel length; and Φ is the velocity potential ($\vec{V} = \nabla\Phi$), G the compressibility, and κ the gas specific heat ratio.

Boundary Conditions

The boundary conditions are surface no-penetration condition, Kutta condition, infinity condition, and wake kinematic and dynamic conditions. They are described as follows:

$$\vec{V}\cdot\vec{n}_g = 0 \qquad \text{on} \quad g(\vec{r}) = 0 \tag{4}$$

$$\Delta C_p|_{sp} = 0 \tag{5}$$

$$\nabla\Phi \to 0 \quad \text{away from} \quad g(\vec{r}) = 0 \quad \text{and} \quad w(\vec{r}) = 0 \tag{6}$$

$$\vec{V} \cdot \vec{n}_w = 0 \qquad \text{on} \quad w(\vec{r}) = 0 \tag{7}$$

and

$$\Delta C_p = 0 \qquad \text{on} \quad w(\vec{r}) = 0 \tag{8}$$

where $\vec{n}_g$ is the unit normal vector of the wing (or a general body) surface, $g(\vec{r}) = 0$; C_p is the surface pressure coefficient; the subscripts sp refers to the edges of separation; and $w(\vec{r}) = 0$ is wake surface(s).

IE Solution

By using the Green's theorem, the integral equation solution of Eq. (1) in terms of the velocity field is given by

$$\begin{aligned}\vec{V}(x,y,z) = {} & \vec{V}_\infty \\ & + \frac{1}{4\pi}\int\int_g \frac{q_g(\xi,\eta,\zeta)}{d^2}\vec{e}_d ds(\xi,\eta,\zeta) \\ & + \frac{1}{4\pi}\int\int_g \frac{\vec{\gamma}_g(\xi,\eta,\zeta)\times\vec{d}}{d^3} ds(\xi,\eta,\zeta) \\ & + \frac{1}{4\pi}\sum_{nw=1}^{NW}\int\int_w \frac{\vec{\gamma}_w(\xi,\eta,\zeta)\times\vec{d}}{d^3} ds(\xi,\eta,\zeta) \\ & + \frac{1}{4\pi}\int\int\int_V \frac{G(\xi,\eta,\zeta)}{d^2}\vec{e}_d d\xi d\eta d\zeta \\ & + \frac{1}{4\pi}\int\int_S \frac{q_S(\xi,\eta,\zeta)}{d^2}\vec{e}_d ds(\xi,\eta,\zeta)\cdot\end{aligned} \tag{9}$$

where $\vec{V}_\infty$ is the free-stream velocity; q is the surface source distribution; $\vec{\gamma}$ is the surface vorticity distribution; the subscript, S, refers to the shock surface; NW is the total number of edges of separation; ds is the infinitesimal surface area; the vector $\vec{d}$ is given by $\vec{d} = (x-\xi)\vec{i}+(y-\eta)\vec{j}+(z-\zeta)\vec{k}$; and $\vec{e}_d$ is defined by $\vec{e}_d = \vec{d}/|\vec{d}|$. It can be seen that the infinity condition, Eq. (6), is automatically satisfied by the integral equation solution.

COMPUTATIONAL SCHEME

Discretisation

In the present paper, three numerical examples (incompressible, high- subsonic and transonic flows as mentioned early) are considered. To simplify the problem, the computations are made for symmetric (no-lifting) flows only. As a consequence, the second and the third integrals in Eq. (9) are not used. In this computational model, the wing surface is represented by a number of rectangular source panels. A uniform rectangular parallelopiped type of volume elements are used throughout the flow field. A constant surface and volume source (q and G) distributions are used over small wing and shock surface panels and over small volume elements. The discretized integral equation solution then becomes

$$\begin{aligned}\vec{V}(x,y,z) &= \vec{V}_\infty \\ &+ \frac{1}{4\pi}\sum_{i=1}^{LG}\sum_{k=1}^{NG} q_{g_{i,k}} \int\int_{g_{i,k}} \frac{1}{d^2}\vec{e}_d ds(\xi,\eta,\zeta) \\ &+ \frac{1}{4\pi}\sum_{i=1}^{LV}\sum_{j=1}^{MV}\sum_{k=1}^{NV} G_{i,j,k} \int\int\int_{V_{i,j,k}} \frac{1}{d^2}\vec{e}_d d\xi d\eta d\zeta \\ &+ \frac{1}{4\pi}\sum_{j=1}^{MS}\sum_{k=1}^{NS} q_{S_{j,k}} \int\int_{S_{j,k}} \frac{1}{d^2}\vec{e}_d ds(\xi,\eta,\zeta)\end{aligned} \tag{10}$$

where the indices, i, j and k refer to the surface panels and field elements; $LG \times NG$ is the total number of wing surface panels; $LV \times MV \times NV$ is the total number of field elements; and $MS \times NS$ is the total number of shock surface panels.

Iterative Scheme

Due to the nature of the non-linearity of high-subsonic and transonic flows, the solutions are obtained through an iterative procedure, where the wing surface source strength and the compressibility are updated through each iteration. The solution procedure follows the successful form of Ref. 10 of two-dimensional computations. Here only the treatment of shocks for transonic flows and their associate mixed nature of flows are described.

Type-Difference Scheme

Transonic flows are characterized by the presence of both subsonic and supersonic regions within the flow field simultaneously. Therefore, transonic flows are described by a mixed elliptic-hyperbolic partial differential equation with the boundary between them unknown apriori. To bo consistent with the mixed nature of the transonic flow, the Murman-Cole type-difference scheme is applied to the present IE computation for $\nabla\rho$ calculations. For subsonic points where the local Mach number is less than unit, central-differencing is used. For supersonic points where the local Mach number is greater than unit, backward-(upstream-) differencing is used. One exception is to use forward-(downstream-) differencing at the first elements after the shock discontinuity. This type-difference scheme is consistent with the nature of the transonic flow, because the local disturbance in a subsonic flow propagates in all directions while in a supersonic flow the local disturbance is confined to the downstream Mach cone of the disturbance.

Shock-Fitting Technique

It should be mentioned that mathematically the fourth (volume) integral term of Eq. (9) includes all compressibility effects including shock discontinuity. Since a relative coarse grid (volume elements) has been used in the present IE computational domain, the contribution of the shock discontinuity is extracted from the forth integral term and it is represented explicitly by the fifth (surface) integral term of Eq. (9). The strength of shock panels, q_S,is equal to the difference of normal velocity across the shock. This can be shown by integrating Eq. (1) over an infinitesimal volume around an infinitesimal area of the shock surface and applying the divergence theorem, one gets

$$\Delta(\nabla\Phi) = V_{2n} - V_{1n} = G\epsilon \tag{11}$$

where ϵ is the infinitesimal thickness normal to the shock surface. By letting $G\epsilon = q_S$ and using Rankine-Hugoniot relation, one finally obtains

$$q_S = [\frac{(\kappa-1)M_{1n}^2+2}{(\kappa+1)M_{1n}^2} - 1]V_{1n} \tag{12}$$

where the subscripts 1 and 2 refer to the conditions ahead and behind of the shock, respectively; and the subscript n refers to the normal component to the shock. The technique to introduce the shock panel and to fit the shock is called shock-fitting technique. In the present calculations for the simplicity, the less accurate, lumped shock panels are used and these shock panels are placed at interface of two adjacent volume elements where the local Mach number changes from greater than unit to smaller than unit.

NUMERICAL EXAMPLES

The presesnt scheme has been applied to a rectangular wing of aspect ratio of 3 with a 5%-thick symmetric circular-arc section. The half-span of the wing (including upper and lower surfaces) is divided into 40×6 quadrilateral panels. The one-half of the computational domain is divided into $40 \times 16 \times 9$ field volume elements in x, y and z directions, respectively. Only symmetric flows with zero angle of attack (non-lifting flows) are considered, and attached flow assumption is made also.

The first numerical example is made for the flow at zero Mach number, an incompressible flow, to validate the computational code. The calculated wing surface pressure distributions are presented in Figure 1. Due to the lack of experimental data for this case, a 2D result calculated by a 2D panel method computational code[19] is plotted also in the figure. The present 3D result at root section (0% semi-span station) compares well with the 2D result.

The second numerical example is made for the flow at free-stream Mach number of 0.7, a shock-free flow. The present calculated surface pressure distributions are presented in Figure 2, along with the experimental data of Ref. 20. In Ref. 20, experimental investigations were made in the supersonic wind tunnel at NASA Ames Research Center. In this paper, the calculated wing surface pressure distributions are compared with the experimental data of Ref. 20 at three span stations located at 0%, 50% and 90% semi-span stations. As the figure shows, the calculated pressure distributions are in qualitative agreement with experimental data, but the calculated pressure level are slightly lower than experimental data at 50% semi-span station. This difference may caused by an effective increase in model thickness due to the boundary layer in the experimental test for which the Reynolds number is 4.0×10^6, as mentioned in Ref.21. The big difference between the calculated values and the experimental data is also observed at the 90% semi-span station, as shown in Figure 2. This difference is mainly due to the fact that in the present computational model the wing tip region is tapered into zero thickness starting at 83% semi-span station and therefore it accelerates the flow in the spanwise direction. In fact the calculated pressure distribution plotted at 90% semi-span station is made at 92% semi-span station where the computational wing model has about zero thickness. Moreover, the difference between the calculated values and the experimental data at 90% semi-span station becomes smaller when the tapering of the computational model wing section thickness starts at 92% semi-span station (the results are not shown here). Although the calculated pressure distributions are not in very good agreement with the experimental data, they are in very good agreement with the computational results obtained by SOUSSA[21] panel method code (see Figure 4 of Ref.21) and by LTRAN3[22] TSD finite- difference method code (see Figure 2 of Ref.22) at 0% ,50% and 70% semi-span station (the comparison is not shown here).

The third numerical example is made for a transonic flow with shock at a free-stream Mach number of 0.9. The calculated pressure distribution compared with the experimental data of Ref.20 at 0%, 50% and 90% semi-span stations, as shown in Figure 3. The set of curves at 50% semi-span station compares well, with the exception that the level of the calculated pressure distribution and the shock strength are lower than the experiment. This is partly due to an effective increase in model thickness in the experiment for which the Reynolds number is 4.4×10^6 as mentioned before, and partly due to the use of the less accurate, or over-simplified, lumped shock panels (in fact they are source points) in the present transonic flow calculation. The difference found at 0% semi-span station can be attributed mainly to the effects of the boundary layer on the wall where the test model was mounted (see Figure 1(a) of Ref. 20).

CONCLUDING REMARKS

An integral equation scheme based on the full-potential equation formulation for transonic flows has been developed. The scheme is capable of handling flows around general three dimensional configurations, although only simple cases are tested in the present paper. The calculated wing surface pressure distribution is reasonably correct including the location and the strength of the shock when using the present simplified computational model. The transonic flow computational results may be improved by using more accurate, distributed shock panels and putting the shock panels in more accurate locations with correct orientations using the Rankine-Hugoniot relations like what was done for airfoil calculations[10]. For the present shock-free flow case, only 6 iterations are used to get a convergent solution, while for the present transonic flow case, 11 iterations are used. The present IEM is effective in terms of the number of iterations compared with those of FDM and FVM, althought the computational cost per IE iteration is more expensive than those of FDM and FVM. Currently, the work has been focused on the improvement of the transonic flow calculations, and in the near furture effects of lifting, unsteadiness and separations will be considered.

ACKNOWLEDGEMENTS

This work has been supported by NASA Langley Research Center under the Grant No.: NAG-1-1170. Dr. Carson Yates, Jr. is the technical monitor. The discussions with Dr. Carson Yates, Jr. are very helpful during the entire course of this research and are gratefully acknowledged.

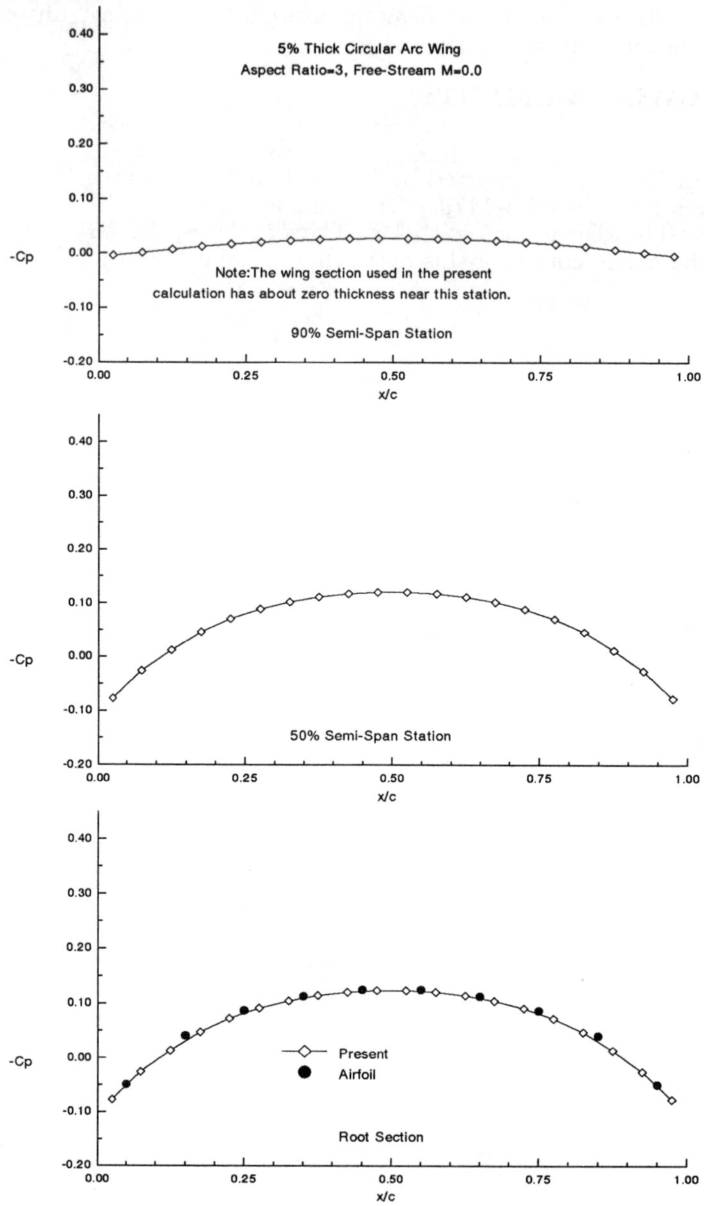

Figure 1. Pressure coefficients for an incompressible flow.

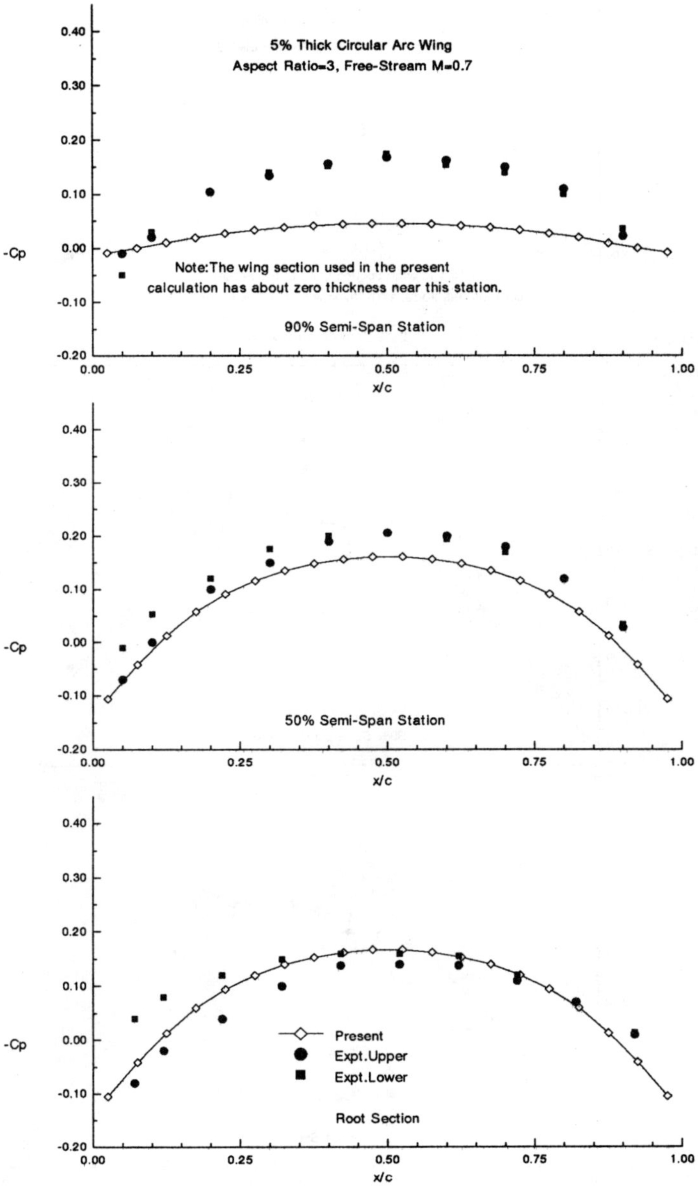

5% Thick Circular Arc Wing
Aspect Ratio=3, Free-Stream M=0.7

Note: The wing section used in the present calculation has about zero thickness near this station.

90% Semi-Span Station

50% Semi-Span Station

Present
Expt.Upper
Expt.Lower

Root Section

Figure 2. Pressure coefficients for a high-subsonic flow.

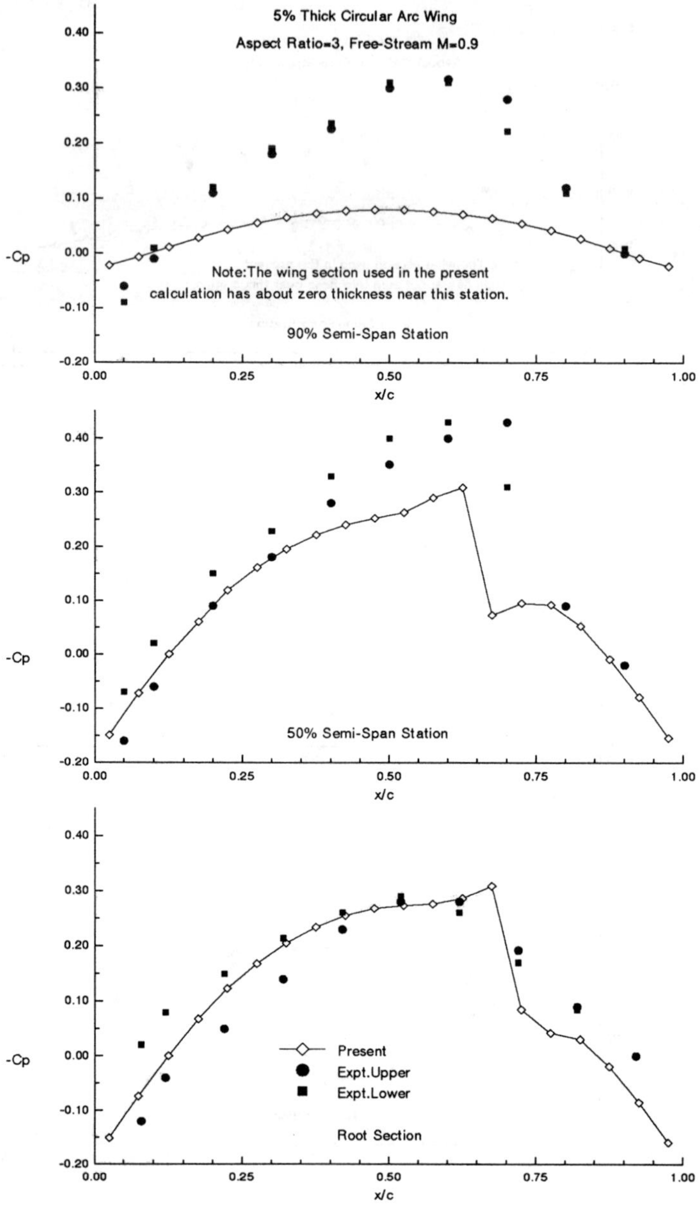

Figure 3. Pressure coefficients for a transonic flow.

REFERENCES

1. Piers, W. J. and Sloof, J. W. Calculation of transonic flow by means of a shock-capturing field panel method, AIAA Paper 79-1459, 1979.

2. Hounjet, M. H. L. Transonic panel method to determine loads on oscillating airfoils with shocks, *AIAA Journal*, 1981, Vol. 19, No. 5, pp.559-566.

3. Tseng, K. and Morino, L. Nonlinear Green's function methods for unsteady transonic flows, *Transonic Aerodynamics*, (Ed. Nixon, D), 1982, pp.565-603.

4. Oskam, B. Transonic panel method for the full potential equation applied to multicomponent airfoils, *AIAA Journal*, 1985, Vol. 23, No. 9, pp.1327-1334.

5. Erickson, L. L. and Strande, S. M. A theoretical basis for extending surface-paneling methods to transonic flow, *AIAA Journal*, 1985, Vol.23, No. 12, pp.1860-1867.

6. Sinclair, P. M. An exact integral (field panel) method for the calculation of two-dimensional transonic potential flow around complex configurations, *Aeronautical Journal*, June/July 1986, Vol. 90, No. 896, pp.227-236.

7. Kandil, O. A. and Yates, E. C., Jr. Computation of transonic vortex flows past delta wings-integral equation approach, *AIAA Journal*, 1986, Vol. 24, No. 11, pp.1729-1736.

8. Madson, M. D. Transonic analysis of the F-16A with under-wing fuel tanks: an application of the TranAir pull-potential code, AIAA Paper 87-1198, 1987.

9. Kandil, O. A. and Hu, H. Transonic airfoil computation using the integral equation with and without embedded Euler domains, *Boundary Elements IX*, Vol. 3, (Eds.: Brebbia, C. A., Wendland, W. L. and Kuhn, G.), Computational Mechanics Publications, Springer-Verlag, 1987, pp. 553-566.

10. Kandil, O. A. and Hu, H. Full-potential integral solution for transonic flows with and without embedded Euler domains, *AIAA Journal*, 1988, Vol. 26, No. 9, pp. 1074-1086.

11. Chu, L-C. Integral equation solution of the full potential equation for three-dimensional, steady, transonic wing flows, Ph.D. dissertation, Dept. of Mechanical Engineering and Mechanics, Old Dominion University, Norfolk, Virginia, March 1988.

12. Sinclair, P. M. A three-dimensional field-integral method for the calculation of transonic flow on complex configurations - theory and preliminary results, *Aeronautical Journal*, June/July 1988, Vol. 92, No. 916, pp. 235-241.

13. Kandil, O. A. and Hu, H. Unsteady transonic airfoil computation using the integral solution of full-potential equation, *Boundary Elements X*, Vol. 2, (Ed.: Brebbia, C. A.), Computational Mechanics Publications, Springer-Verlag, 1988, pp. 352-371.

14. Kandil, O. A. and Hu, H. Integral solution of unsteady full- potential equation for a transonic pitching airfoil, *Journal of Aircraft*, 1990, Vol. 27, No. 2, pp. 123-130.

15. Ogana, W. Transonic integro-differential and integral equations with artificial viscosity, *Engineering Analysis with Boundary Elements*, 1989, Vol. 6, No. 3, pp. 129-135.

16. Hu. H. Field-boundary element solution for a transonic pitching wing, *Topics in Engineering, Vol. 7: Advances in Boundary Elements Methods in Japan and USA*, (Eds.: Tanaka, M., Brebbia, C. A. and Shaw, R.), Computational Mechanics Publication, 1990, pp. 295-305.

17. Hu, H. Field-boundary element computation for transonic flows, *Computation Engineering with Boundary Elements*, Vol. 1, (Eds.: Grilli, S., Brebbia, C. A. and Cheng, A. H-D.), Computational Mechanics Publication, 1990, pp. 319-330.

18. Hu, H. Unsteady transonic wing flow computations using field- boundary element methods, to appear in the *Journal, Engineering Analysis with Boundary Elements.*

19. Hu, H. and Simms, A. L. (work in progress.)

20. Lessing, H. C., Troutman, J. L. and Menees, G. P. Experimental determination of the pressure distribution on a rectangular wing oscillating in the first bending mode for Mach numbers from 0.24 to 1.30, NASA TN D-344, 1960.

21. Yates, E . C. Jr., Cunningham, H. J. and Desmarais, R. N. Subsonic aerodynamic and Flutter characteristics of several wings calculated by the SOUSSA P1.1 panel method, AIAA Paper 82-0727, 1982.

22. Guruswamy, P. and Goorjian, P. M. Comparisons between computations and experimental data in unsteady three-dimensional transonic aerodynamics, including aeroelastic applications, AIAA Paper 82-0690-CP, 1982.

Solution of the Transonic Integral Equation Using Discontinuous Linear and Quadratic Elements

W. Ogana

Department of Mathematics, University of Nairobi, P.O. Box 30197, Nairobi, Kenya

INTRODUCTION

The transonic integral equation has been solved predominantly by numerical methods which assume that the dependent variable is constant within a disretization element. In order to improve accuracy, Nixon [1] assumed that within the discretization element the dependent variable was constant in the streamwise direction but varied linearly in the transverse direction. Ogana [2-7] has made a similar assumption but has formulated the problem in more general terms using boundary element methods. Consequently he has been able to apply linear and quadratic variation, in the transverse direction, giving rise to "hybrid" elements for both the integral and integro-differential equations. These methods yield improved accuracy but they cannot be conveniently applied to a large computational domain without a sharp increase in nodes and discretization elements.

The transverse integral is over a semi-infinite interval and may be evaluated analytically on applying the decay function due to Oswatitsch [8-10]. If numerical evaluation of the transverse integral is preferred then the decay function due to Ogana [4,6,11] should be used since it extends the solution to points in front and aft of the airfoil, unlike the Oswatitsch approximation which restricts attention to the airfoil surface. It is desirable to provide a formulation which permits linear and higher order variations of the dependent variable, in the steamwise direction, but treats the transverse semi-infinite integral by a more general quadrature scheme that is capable of extending the computational domain without a large increase in nodes. Since we are usually interested in values on the airfoil surface, it is preferrable to use a method, such as Radau quadrature, which automatically places nodes on the airfoil surface.

In this paper we develop a method of solving the transonic integral equation by assuming that the dependent variable has discontinous linear and quadratic variations, in the streamwise direction, within a discretization element. The transverse integral is evaluated by Radau quadrature. The method is tested on biconvex and NACA0012 airfoils in nonlifting high subsonic flows, using the decay function of Ogana [4,6,11]. Results are compared with finite difference solutions.

INTEGRAL EQUATION AND BOUNDARY ELEMENTS

For nonlifting flows, the small-disturbance transonic integral equation may be written

$$u(x,y)=u_T(x,y)+\upsilon u^2(x,y)+\int_0^{\infty}\int_{-\infty}^{\infty}K(\xi-x,\zeta-y)u^2(\xi,\zeta)\delta\xi\delta\zeta \qquad (1)$$

where u is the perturbation streamwise velocity component and υ is dependent on the height to width ratio of the cavity which surrounds the singularity [12]. The kernel K is

$$K(\xi-x,\zeta-y) = \frac{(\zeta-y)^2-(\xi-x)^2}{2\pi\left[(\xi-x)^2+(\zeta-y)^2\right]^2} \qquad (2)$$

The quantity u_T is the contribution to the perturbation velocity due to the thickness distribution and is given by

$$u_T(x,y) = -\frac{(\gamma+1)}{\pi K_s^{3/2}}\int_0^1 \frac{(\xi-x)F'(\xi)}{(\xi-x)^2+y^2}\,d\zeta \qquad (3)$$

in which γ is the ratio of specific heats, K_s is the transonic similarity parameter and F(x) the thickness distribution.

For numerical approximation of Eq.(1), first divide the computational domain into M strips parallel to the y-axis, labelled $\Delta_1,\Delta_2,...,\Delta_m$. Let the strip Δ_j have width $2\delta_j$ and centre with x-coordinate, $x = X_j$. In order to extend the computational domain in the streamwise direction without using many strips, wider strips may be chosen away from the airfoil. Let there be N global nodes $Q_1, Q_2, \ldots, Q_N$ at which the solution is required and let the coordinates of Q_i be (x_i, y_i). Apply Eq.(1) at the node Q_i to find

$$u_i = u_{Ti} + \sum_{j=1}^{m} I_j, \qquad i=1,2,\ldots,N \tag{4}$$

where $u_i \equiv u(Q_i)$, $u_{Ti} \equiv u_T(Q_i)$ and

$$I_j = \iint_{\Delta_j} Ku^2 ds = \int_0^{\infty} \int_{x_j-\delta_j}^{x_j+\delta_j} K(\xi - x_i, \zeta - y_i) u^2(\xi, \zeta) d\xi d\zeta \tag{5}$$

Within Δ_j define a local coordinate, t, in the streamwise direction, by

$$t = (\xi - x_j)/\delta_j \tag{6}$$

Hence

$$I_j = \delta_j \int_0^{\infty} \int_{-1}^{1} K(t\delta_j + X_j - x_i, \zeta - y_i) u^2(t, \zeta) dt d\zeta \tag{7}$$

The numerical method for solving Eq.(4) depends on how $u^2(t,\zeta)$ is assumed to vary in the streamwise direction, within Δ_j. Existing methods are based on the assumption that u^2 is constant within Δ_j, in the streamwise direction . For the current formulation, we assume that $u^2(t,\zeta)$ has discontinuous linear and quadratic variation, in the streamwise direction, within Δ_j, i.e for $-1 \le t \le 1$. Consequently in Δ_j there will be two columns of nodes, for linear variation (Fig.1a) and three columns for quadratic variation (Fig.1b) . The number of nodes per column depends on the quadrature formula used to evaluate the semi-infinite transverse integral. When evaluating quantities within Δ_j these nodes will be referred to as local nodes to distinguish them from the global node Q_i at which the solution is sought.

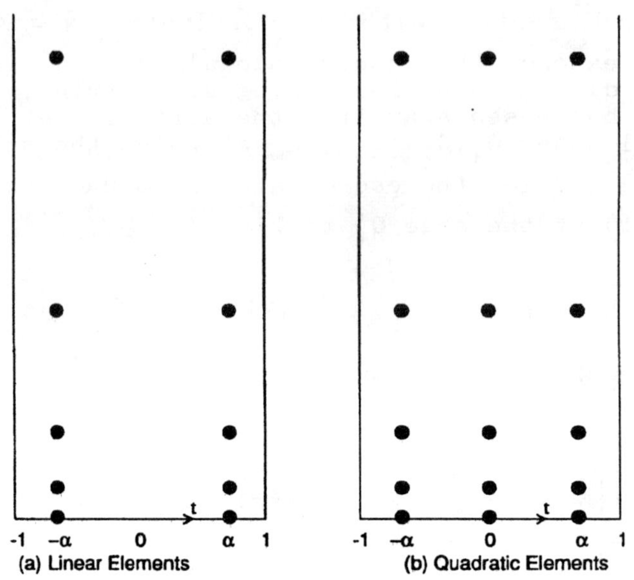

Fig. 1 Local coordinates and distribution of nodes in discretization elements

Discontinuous Linear Elements

For discontinuous linear variation, following Dyka and Millwater [13] , for any $t \in [-1,1]$ we may write

$$u^2(t,\zeta) = \theta_1 u^2(t_1,\zeta) + \theta_2 u^2(t_2,\zeta) \tag{8}$$

where

$$t_1 = -\alpha \quad , \quad t_2 = \alpha \tag{9}$$

while $u(t_1,\zeta)$ and $u(t_2,\zeta)$ are the values of $u(t,\zeta)$ along the first and second columns, respectively (Fig.1a). The first column is located on the line $t = -\alpha$ or , from Eq.(5), the line $x = X_j - \alpha\delta_j$ while the second column is on $t = \alpha$ or $x = X_j + \alpha\delta_j$. The shape functions in Eq.(8) are

$$\theta_1(t) = \frac{1}{2}(1 - \frac{t}{\alpha}), \quad \theta_2(t) = \frac{1}{2}(1 + \frac{t}{\alpha}), \tag{10}$$

in which $0 < \alpha \leq 1$. The choice $\alpha = 1$ corresponds to continuous linear variation while $\alpha \neq 1$ corresponds to discontinuous linear variation. Substitution of Eq.(8) into Eq.(7) and integration with. respect to t yields

$$I_j = \int_0^{\infty} \left[\sum_{k=1}^{2} W_k(p_1,p_2,\zeta-y_i)u^2(t_k,\zeta) \right] d\zeta \tag{11}$$

where

$$W_1(p_1,p_2,\zeta-y_i) = \frac{1}{4\pi\alpha}\left[-(h+\alpha)G_1 + \frac{1}{\delta_j} G_2\right] \tag{12a}$$

$$W_2(p_1,p_2,\zeta-y_i) = \frac{1}{4\pi\alpha}\left[(h-\alpha)G_1 - \frac{1}{\delta_j} G_2\right] \tag{12b}$$

in which

$$h = (X_j - x_i)/\delta_j \tag{13a}$$

$$G_1 = \frac{p_1}{p_1^2+(\zeta-y_i)^2} - \frac{p_2}{p_2^2+(\zeta-y_i)^2} \tag{13b}$$

$$G_2 = \frac{p_1}{p_1^2+(\zeta-y_i)^2} - \frac{p_2}{p_2^2+(\zeta-y_i)^2} + \frac{1}{2}\ln\left[\frac{p_2^2+(\zeta-y_i)^2}{p_1^2+(\zeta-y_i)^2}\right] \tag{13c}$$

$$p_1 = X_j - x_i - \delta_j \quad , \qquad p_2 = X_j - x_i + \delta_j \tag{13d}$$

If the global node $Q_i \notin \Delta_j$ then $|X_j - x_i| > \delta_j$ and Eq.(13) shows $p_1 \neq 0$ and $p_2 \neq 0$. Hence I_j is nonsingular. Suppose, however, that Q_i is located within the first column of local nodes in Δ_j, then $x_i = X_j - \alpha\delta_j$ and

$$P_1 = -(1 - \alpha)\delta_j, \quad P_2 = (1 + \alpha)\delta_j \tag{14a}$$

This equation shows that if $\alpha = 1$, i.e. continuous linear variation is used, then $p_1 = 0$, leading to singularities in I_j at $\zeta = y_i$. Similarly if Q_i is located within the second column of local nodes then $x_i = X_j + \alpha\delta_j$ and

$$p_1 = -(1 + \alpha)\delta_j, \quad p_2 = (1 - \alpha)\delta_j \tag{14b}$$

Hence the use of the continous linear variation gives

$p_2 = 0$, leading to singularities in I_j at $\zeta = y_i$. We note, however, that if $\alpha \neq 1$, i.e. discontinuous linear variation is used, then $p_1 \neq 0$ and $p_2 \neq 0$, for all X_j and x_i, and the integral I_j contains no singularities.

Discontinuous Quadratic Elements

For discontinuous quadratic variation, following Dyka and Millwater [13], for any $t \in [-1, 1]$ we may write

$$u^2(t,\zeta) = \theta_1 u^2(t_1,\zeta) + \theta_2 u^2(t_2,\zeta) + \theta_3 u^2(t_3,\zeta) \tag{15}$$

where

$$t_1 = -\alpha, \qquad t_2 = \alpha, \qquad t_3 = 0 \tag{16}$$

while $u(t_1,\zeta)$, $u(t_2,\zeta)$ and $u(t_3,\zeta)$ are the values of $u(t,\zeta)$ along the first, second and third columns, respectively (Fig. 1b). The first column is located along the line $t = -\alpha$ or $x = X_j - \alpha\delta_j$, the second along the line $t = \alpha$ or $x = X_j + \alpha\delta_j$, and the third along the line $t = 0$ or $x = X_j$. The shape functions in Eq.(15) are

$$\theta_1 = \frac{t}{2\alpha}\left[\frac{t}{\alpha} - 1\right], \quad \theta_2 = \frac{t}{2\alpha}\left[\frac{t}{\alpha} + 1\right], \quad \theta_3 = \left[1 - \frac{t}{\alpha}\right]\left[1 + \frac{t}{\alpha}\right] \tag{17}$$

in which $0 < \alpha \leq 1$. The choice $\alpha = 1$ corresponds to continuous quadratic variation while $\alpha \neq 1$ corresponds to discontinuous quadratic variation.

Substitution of Eq.(17) into Eq.(7) and integration with respect to t yields

$$I_j = \int_0^\infty \left[\sum_{k=1}^{3} W_k(p_1, p_2, \zeta - y_i)\, u^2(t_k, \zeta)\right] d\zeta \tag{18}$$

where

$$W_1(p_1, p_2, \zeta - y_i) = \frac{1}{4\pi\alpha^2}\left[-h(h+\alpha)G_1 + \frac{(2h+\alpha)}{\delta_j} G_2 - \frac{1}{\delta_j^2} G_3\right] \tag{19a}$$

$$W_2(p_1, p_2, \zeta - y_i) = \frac{1}{4\pi\alpha^2}\left[-h(h-\alpha)G_1 + \frac{(2h-\alpha)}{\delta_j} G_2 - \frac{1}{\delta_j^2} G_3\right] \tag{19b}$$

$$W_3(p_1,p_2,\zeta-y_i) = \frac{1}{2\pi\alpha^2}\left[(h^2-\alpha^2)G_1-\frac{2}{\delta_j}G_2+\frac{1}{\delta_j^2}G_3\right] \quad (19c)$$

in which h, G_1, G_2, p_1, and p_2 are given by Eq.(13) while

$$G_3 = \frac{p_1^3}{p_1^2+(\zeta-y_i)^2} - \frac{p_2^3}{p_2^2+(\zeta-y_i)^2} + 4\delta_j - 2(\zeta-y_i)\tan^{-1}\left[\frac{2\delta_j(\zeta-y_i)}{p_1p_2+(\zeta-y_i)^2}\right] \quad (20)$$

Using arguments analogous to those for linear elements, it can readily be shown that the use continuous quadratic elements leads to singular integrals while the use of discontinuous quadratic elements leads to non-singular integrals. The absence of singularities makes discontinuous elements (linear and quadratic) more attractive to apply even though they lead to slightly more nodes than the corresponding continuous elements.

RADAU QUADRATURE AND DECAY FUNCTION

Equations (11) and (18) contain integrals of the form

$$I_{jk} = \int_0^\infty W_k(p_1,p_2,\zeta-y_i)u^2(t_k,\zeta)\,d\zeta \quad (21)$$

for values of k, t_k and W_k depending on whether linear or quadratic elements are used. Such integrals may be approximated by suitable quadrature schemes so that Eq.(4) reduces to a nonlinear algebraic system. Since we are usually interested in values on the airfoil surface, it is appropriate to apply a method, such as Radau quadrature, which automatically places nodes on the airfoil, namely at $\zeta = 0$. To apply Radau quadrature, first transform the transverse coordinate, ζ, through

$$s = \frac{\zeta-1}{\zeta+1} \quad \Rightarrow \quad \zeta = \frac{s+1}{1-s} \quad (22)$$

and substitute in Eq.(21) to obtain

$$I_{jk} = \int_{-1}^{1} f(s)ds \tag{23}$$

where

$$f(s) = \frac{2W_k[p_1,p_2,\zeta(s)-y_i]u^2[t_k,\zeta(s)]}{(1-s)^2} \tag{24}$$

Application of an n-point Radau quadrature formula to Eq.(23) yields, see for instance [14]

$$I_{jk} = \sum_{l=1}^{n} a_l f(s_l) \tag{25}$$

where n is the number of quadrature points and s_l and a_l are the quadrature abscissas and weights, respectively, with the abscissa $s_1 = -1$ fixed. The method can be made exact for polynomials of degree up to 2(n-1). Substitution of Eqs.(23) to (25) into Eq.(11) or Eq.(18) yields

$$I_j = \sum_k \sum_{l=1}^{n} h_{ij}^{kl} \left(u_j^{kl}\right)^2 \tag{26}$$

where the summation is over k = 1,2 for linear elements and k = 1,2,3 for quadratic elements. The quantity

$$u_j^{kl} = u(t_k,\zeta_l) \tag{27}$$

in which $\zeta_l = \zeta(s_l)$, is the value of u at the quadrature point l in column k within Δ_j. The influence coefficient h_{ij}^{kl} is given by

$$h_{ij}^{kl} = \frac{2a_l W_k(p_1,p_2,\zeta_l - y_i)}{(1-s_l)^2} \tag{28}$$

and defines the interaction between the global node Q_i and the quadrature point l located in column k within Δ_j. Substituting Eq.(26) into Eq.(4) and identifying the quantity u_j^{kl} with the appropriate global node leads to the nonlinear system

$$u_i = u_{Ti} + \sum_{j=1}^{N} b_{ij} u_j^2 , \qquad i=1,2,\ldots,N \tag{29}$$

where $u_j \equiv (Q_j)$ and $\{b_{ij}\}$ is a dense N x N matrix which depends on how the local quadrature points are identified as global nodes.

Use of Eq.(26) leads to a direct method which determines the solutions at nodes on the airfoil and throughout the domain. The matrix elements are computed once and stored. Depending on the number of nodes per row and points in Radau quadrature, this formulation could lead to a large system. To reduce the size of the system, we may solve for nodes on the x-axis only and approximate velocities off the x-axis by a decay function. The Oswatitsch decay function [8-10] should be used if attention is restricted entirely to the airfoil surface while the Ogana decay function [4,6,11] should be used if computation extends in front and aft of the airfoil. In this paper we apply the latter method. For nonlifting flows, the Ogana decay function is

$$u(x,y) = \frac{u(x,0)}{1+c(x)y+b(x)y^2} \tag{30}$$

where

$$c(x) = \begin{cases} Y'(x)/u , & 0 \le x \le 1 \\ 0, & \text{otherwise} \end{cases} \tag{31a}$$

$$b(x) = \left| c^2(x) - \frac{1}{2u} \left[(u - 1)u_{xx} + u_x^2 \right] \right| \tag{31b}$$

in which Y(x) is the modified airfoil profile. The velocity u and its derivatives are evaluated at (x,0). For quadrature points off the x-axis, using Eq.(30) in Eq.(27) and substituting the result in Eq.(26) yields

$$I_j = \sum_k h_{ij}^{k1} \left(u_j^{k1} \right)^2 \tag{32}$$

where the summation is over k = 1,2 for linear elements and k = 1,2,3 for quadratic elements. From Eq.(27) we conclude that $u_j^{k1} \equiv u(t_k,0)$ since the first quadrature point, corresponding to $\iota = 1$, is on the x-axis .The influence coefficient h_{ij}^{k1} is given by

$$h_{ij}^{k1} = \sum_{l=1}^{n} \frac{2a_l W_k(p_1, p_2, \zeta_l - y_i)}{\left[(1-s_l)^2(1+c_k\zeta_l+b_k\zeta_l^2)\right]^2} \qquad (33)$$

where $c_k = c(t_k)$ and $b_k = b(t_k)$. It defines the interaction between the global node Q_i, on the x-axis, with the first quadrature node, also on the x-axis, contained in column k in Δ_j.

Substituting Eq.(32) into into Eq.(4) and identifying u_j^{k1} with the appropriate node on the airfoil leads to a nonlinear algebraic system similar to Eq (29) except that N now stands for the number of nodes on the x-axis. Because the nodes are confined to the x-axis, this approach leads to a much smaller system than the direct method. Unlike the direct method, the quantities h_{ij}^{k1}, hence the matrix elements b_{ij}, have to be evaluated at each iteration since they depend on c(x) and b(x) which vary with the dependent variable u(x,0).

RESULTS AND DISCUSSION

The discontinous linear and quadratic elements described here were tested on airfoils in high subsonic flows, using the Ogana decay function. In calculations involving linear elements the parameter α in Eq.(10) was chosen to be 0.5 for all discretization elements while for quadratic elements α was chosen thus:

$$\alpha = \begin{cases} 0.75, & \Delta_j \text{ next to airfoil edge} \\ 0.5, & \text{Otherwise} \end{cases} \qquad (34)$$

These values of α are not unique and were chosen simply for the convenience of locating the positions of nodes in Δ_j . For NACA0012 airfoil the linearised velocity was corrected near the leading edge as described in Nixon *et al* [9].

Fig. 2 shows the coefficient of pressure plots for a

biconvex airfoil of thickness ratio $\delta = 0.1$ and at

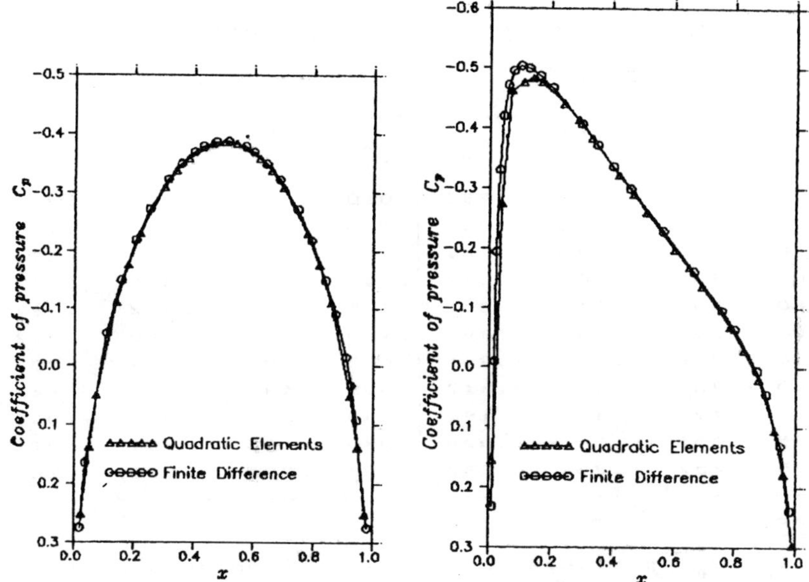

Fig.2 Pressure coefficient for biconvex airfoil $\delta = 0.1$, $M_\infty = 0.7$

Fig.3 Pressure coefficient for NACA0012 airfoil $\delta = 0.12$, $M_\infty = 0.63$

$M_\infty = 0.7$ while Fig.3 shows the plots for a NACA0012 airfoil at $M_\infty = 0.63$. The finite difference results for the biconvex airfoil are obtained from Ballhaus *et al* [15] while the NACA0012 results are obtained from Ballhaus *et al* [16].

Table 1. shows a comparison of the pressure coefficients at selected nodes calculated using quadratic elements, linear elements and finite differences. The boundary element and finite difference results compare favourably except near the airfoil edges where the velocity gradients are large. These characteristics have been displayed by other integral equation methods. The quadratic elements results were computed using 13 discretization elements with 7 on the airfoil and 3 each in front and aft of the airfoil. The linear elements results were obtained using 20 discretization elements with 10 on the airfoil and 5 each in front and aft of the airfoil. In both cases smaller widths for discretization elements were chosen near the airfoil edges while the widths became larger away from the airfoil. Table 2 compares the boundary elements results obtained when the

Table 1. Comparison of pressure coefficients

BICONVEX *Airfoil* $\delta = 1.0$, $M_\infty = 0.7$				**NACA0012** *Airfoil* $\delta = 0.12$, $M_\infty = 0.63$			
x	C_P^{QE}	C_P^{LE}	C_P^{FD}	x	C_P^{QE}	C_P^{LE}	C_P^{FD}
.025	.252	.251	.232	.010	.155	.207	.342
.075	.051	.046	.026	.070	-.461	-.444	-.486
.180	-.177	-.176	-.184	.140	-.481	-.477	-.496
.300	-.309	-.308	-.312	.245	-.440	-.439	-.442
.380	-.358	-.357	-.361	.335	-.382	-.382	-.384
.500	-.386	-.385	-.387	.470	-.290	-.290	-.292
.620	-.358	-.357	-.361	.605	-.198	-.198	-.202
.700	-.309	-.308	-.312	.695	-.137	-.137	-.142
.820	-.177	-.176	-.184	.830	-.029	-.030	-.036
.925	.052	.046	.026	.930	.104	.103	.092
.975	252	.251	.232	.990	.296	.298	.288

Notes:

C_P^{QE}, C_P^{LE}, C_P^{FD} pressure coefficients from quadratic elements, linear elements and finite differences, respectively.

Table 2. Magnitudes of the differences in pressure coefficients between computation on the airfoil only and computation in the extended domain

BICONVEX *Airfoil*				**NACA0012** *Airfoil*			
Quadratic		*Linear*		*Quadratic*		*Linear*	
x	d	x	d	x	d	x	d
.025	6.82	.025	11.50	.010	38.40	.01	96.50
.075	2.11	.125	2.26	.070	3.12	.06	19.40
.180	1.17	.225	1.17	.140	1.69	.15	4.19
.300	1.28	.325	1.40	.245	0.08	.27	1.94
.380	1.17	.425	1.24	.335	0.06	.39	1.20
.500	1.06	.525	1.21	.470	0.05	.51	0.09
.620	1.17	.625	1.30	.605	0.05	.63	0.08
.700	1.28	.725	1.53	.695	0.05	.75	0.08
.820	1.17	.825	1.95	.830	0.06	.87	1.04
.925	2.11	.925	4.00	.930	1.21	.97	6.65
.975	6.82	.975	11.50	.990	10.10	.99	11.90

Notes:

$d \times 10^{-3}$ - Magnitude of pressure coefficient difference

far-field boundary is located about two chord lengths from the airfoil with results obtained when computation is restricted on the airfoil. The two methods yield results of comparable accuracy except in the neighbourhood of the airfoil edges.

REFERENCES

1. Nixon, D., 'Calculation of Transonic Flows Using an Extended Integral Equation Method,' *AIAA.J.*, Vol.15, pp. 295 - 296, 1977.
2. Ogana,W ., 'Boundary Element Methods in Two-Dimensional Transonic Flows,' in *Boundary Elements IX - Vol.3: Fluid Flow and Potential Applications*, (Eds.) C.A.Brebbia, W.L.Wendland and G.Kuhn, Computational Mechanics Publications, Southampton, and Springer-Verlag, Berlin, and New York, pp.567 - 582, 1987.
3. Ogana, W., 'Boundary Element Solution of the Transonic Integral Equation,' *Afrika Matematika*, Vol. 2, pp.5-28, 1988.
4. Ogana, W., 'Boundary Element Solution of the Transonic Integro-differential Equation involving a Decay Function,' *Kenya Journal of Sciences*, Vol. 9, pp.23-28,1988.
5. Ogana, W., 'Boundary Element Solution of the Transonic Integro-differential Equation,' *Engineering Analysis With Boundary Elements*, Vol. 6, pp.170-179, 1989.
6. Og a na, W., 'Solution of the Transonic Integro-differential Equation using a Decay Function,' *Appl. Math. Modelling*, Vol. 14, pp. 30-35, 1990.
7. Ogana, W., 'An a lysis of Transonic Integral Equations: Part II - Boundary Element Methods,' *AIAA J.*, Vol. 28, pp. 556-558, 1990.
8. Spreiter, J. R . and Alksne, A.Y. 'Theoretical Prediction of Pressure Distribution on Nonlifting Airfoils at High Subsonic Speeds,' *NACA Rep.* 1217, 1955.
9. Nixon, D. and Hancock, G.J., 'High Subsonic Flow past a Steady Two-dimensional Aerofoil,' *ARC CP* 1280, 1974.
10. Niyogi, P., *Integral Equation Method in Transonic Flow*, Lecture Notes in Physics, Vol. 157, Springer-Verlag, Berlin and New York, 1982.

11. Ogana, W., 'Choosing the Decay Function in the Transonic Integral Equation,' *Kenya Journal of Sciences*, Vol. 9, pp. 11-21, 1988.

12. Ogana, W. and Spreiter, J.R., 'Derivation of an Integral Equation for Transonic Flows,' *AIAA J.*, Vol. 15, pp. 281-283, 1977.

13. Dyka, C. T. and Millwater, H. R., 'Formulation and Integration of Discontinous and Continous Quadratic Boundary Elements for Two-dimensional Potential and Elastostatics,' *Computers& Structures*, Vol 31, pp. 495 - 504, 1989.

14. Stroud, A. and Secrest, D., *Gaussian Quadrature Formulas*, Prentice-Hall, Englewood Cliffs, N.J., U.S.A, 1966.

15. Ballhaus, W.F., Jameson, A. and Albert, J., 'Implicit Approximate-Factorization Schemes for the Efficient Solution of a Steady Transonic Flow Problems,' *NASA TMX* - 73202, 1977.

16. Ballhaus, W.F., Goorjian, P.M., 'Implicit Finite - Difference Computations of Unsteady Transonic Flows About Airfoils,' *AIAA J.*, Vol. 15, pp. 1728 - 1735, 1977.

Investigation of the Aerodynamic Performance of a Formula 1 Multi-Aerofoil Spoiler using a Second Order Complex Variable Boundary Element Method

T.W. Chiu, B. Wood, D.J. Buckingham

School of Engineering, University of Exeter, Exeter EX4 4QF, U.K.

Abstract

An indirect complex variable boundary element method (CVBEM) has been used to investigate the performance of a Formula-I Multi-Aerofoil Spoiler. A few possible modification to the original design has been looked at and an optimum design was selected. This method is found to be economical in this initial design and optimisation process, which will vastly reduce the amount of wind tunnel work and road-tests required.

1 Introduction

Potential flow problems are a major area in the application of the Boundary Element Method (BEM). While in real life viscous effects exist in most aerodynamic applications, it is generally accepted that the high speed incompressible flow over streamlined bodies can effectively be approximated by potential flow models. This paper presents an investigation and modification of a Formula I spoiler using the Complex Variable Boundary Element Method (CVBEM).

The design of racing car spoiler has conventionally been done through wind tunnel and road testing. While experiments are indispensable for the testing of the design, computer modelling in the early design stage will reduce substantially the amount of experimental work needed to optimise the spoiler configuration.

The ideal spoiler will produce a very strong downward force while keeping the drag force as small as possible. The centre of pressure should at the same time as far aft as possible. The range of Reynolds numbers in car racing is normally lower than in aerospace applications. This requires the investigation to take a different approach from that of aircraft wing design, and highlights the versatility of the CVBEM in incompressible potential flows. The present project is the first stage of a comprehensive investigation which aims at improving the performance of an existing Formula I spoiler composed of 3 aerofoils. For practical reasons it is necessary to keep the alternative configurations of the real spoiler to a minimum. It is economically desirable to modify rather than replace the present design and to keep alterations to a minimum. So a two-dimensional second order CVBEM is employed at this stage to test a large number of possible modified configurations. Since the CVBEM is the most economical numerical scheme available, parameters, such as the relative position of aerofoils, can be varied at small steps to inspect the effects so as to determine the optimum values.

The following sections present the basic formulation and some numerical problems encountered, such as the imposition of the Kutta Condition. Then the results obtained for a few alternative designs will be presented.

2 CVBEM with Vortex Distribution

The CVBEM used in the analysis is an indirect method in which a vortex distribution is assumed over the model surface. A vortex distribution is mathematically identical to a source distribution (single-layer BEM) except that the source strength is *imaginary* [1]. The model boundary is approximated by a number of straight line elements, on which the vortex strength varies linearly from one end to the other. A collocation point is located at the midpoint of each of the elements where the external Neumann boundary condition is imposed. The derivation of this numerical model for general two-dimensional potential flows is described in detail in [1].

As a result of the imposition of the boundary conditions, a system of linear equations is formed:

$$\sum_{j=1}^{m}\Big[-\frac{\gamma_j}{2\pi}\text{Re}\left\{\frac{l_j(z_{p+1}-1)}{(z_j-z_{j+1})(\overline{z_{k+1}-z_k})}\right\}$$
$$-\frac{\gamma_{j+1}}{2\pi}\text{Re}\left\{\frac{l_j(z_p-1)}{(z_j-z_{j+1})(\overline{z_{k+1}-z_k})}\right\}\Big]$$
$$+\text{Im}\left\{\frac{U_\infty}{(\overline{z_{k+1}-z_k})}\right\}=0 \qquad k=1,2,\cdots,m \tag{1}$$

$$z_{p+1}=\frac{z_{k+\frac{1}{2}}-z_{j+1}}{z_j-z_{j+1}}\log\left(\frac{z_{k+\frac{1}{2}}-z_{j+1}}{z_{k+\frac{1}{2}}-z_{j+1}}\right)$$

$$z_p=\frac{z_{k+\frac{1}{2}}-z_j}{z_j-z_{j+1}}\log\left(\frac{z_{k+\frac{1}{2}}-z_{j+1}}{z_{k+\frac{1}{2}}-z_{j+1}}\right)$$

m = number of elements

l_j = length of the j-th element

z_j, z_{j+1} = complex coordinates of the j-th and $(j+1)$-th boundary points, forming the j-th element

$z_{k+\frac{1}{2}}$ = complex coordinate of the collocation point of the $k-$th element $=\frac{1}{2}(z_k+z_{k+1})$

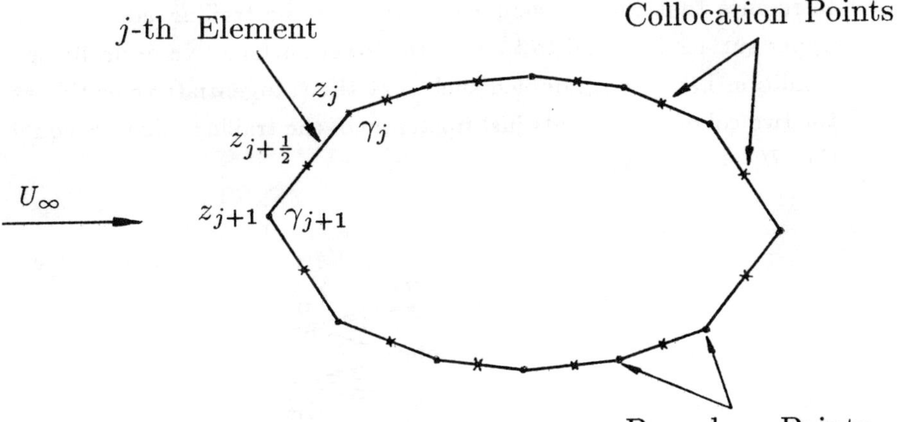

Fig.1 Representation of boundary by linear elements.

The solution of this set of linear equation will give the vortex strengths (γ_j's) at the boundary points (z_j's) (fig.1) which will then be used to calculate the velocity at any point in the flow field.

3 Numerical Difficulties and Kutta Condition

The above formulation with vortex distribution can be used for non-lifting as well as lifting flows. In the case of non-lifting flow, a source distribution is preferred and a set of linear equations similar to equation (1) would be obtained. The matrix thus formed always has dominant diagonal entries and thus the equations could be solved very efficiently using Gauss-Seidel iteration. However, with a vortex distribution, the system of equations hence obtained is singular because it is possible to satisfy the boundary conditions with more than one value of circulation around the body. Therefore the equations cannot be handled by normal solution methods. It is found that if the system of equations is solved by *singular value decomposition* (SVD) [2], a minimal solution will be obtained which corresponds to the case of zero net circulation around the body.

In the case of lifting body, the numerical difficulty is further complicated by the requirement of a Kutta condition. It is required that the fluid leaves the trailing edge smoothly without shear. There are two possible ways to impose this condition:

1. The first way, which is particularly good for cusped trailing edge, is to require that the (tangential) velocity at the trailing edge on the upper surface be equal to that on the lower surface. Numerically the condition has to be modified such that the (tangential) velocities at the two collocation points just upstream of the trailing edge are equal (i.e. $U_{t1} = U_{t2}$, fig.2).

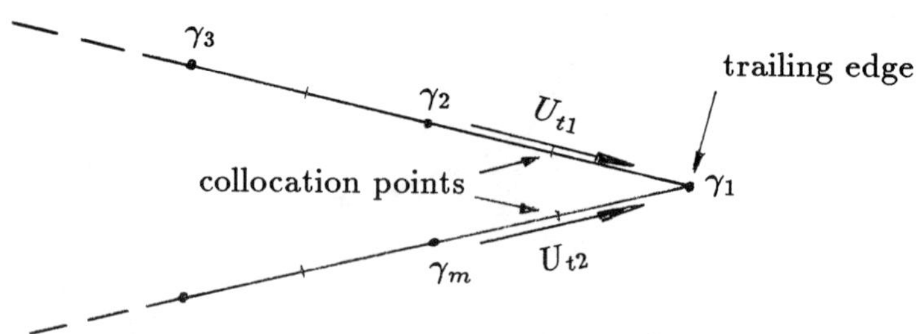

Fig.2 Imposition of Kutta condition at trailing edge.

2. Another way of imposing the Kutta condition is to specify that the vorticity at the trailing edge is zero ($\gamma_1 = 0$).

In any case, the Kutta condition will mean an extra equation to satisfy, i.e. we will have $m+1$ equations for m unknowns. In the first instant, it seems that the system of equations is overdetermined [3]. However, the original set of m equations is underdetermined, and what it lacks is exactly the Kutta condition. So the equations will, in theory, give one unique set of solution for the γ_j's. (Alternatively one can replace any one of the original equations by the Kutta condition equation. This would not affect the result [3].) Again, SVD is employed which can cope with non-square left hand side matrices. It is also found that, with all the aerofoil problems investigated, the above two ways of imposing the Kutta condition (i.e. $U_{t1} = U_{t2}$ or $\gamma_1 = 0$) give practically the same solution.

The use of $\gamma_1 = 0$ as the Kutta condition is in fact easier to formulate and can make the computer implementation slightly more efficient. Suppose the original m equations are represented by

$$[A_{ij}](\gamma_j) = (B_i)$$

$$\Longrightarrow \quad \begin{pmatrix} a_{11} & a_{12} & \cdots & a_{1m} \\ a_{21} & a_{22} & \cdots & a_{2m} \\ \vdots & \vdots & \ddots & \vdots \\ a_{m1} & a_{m2} & \cdots & a_{mm} \end{pmatrix} \begin{pmatrix} \gamma_1 \\ \gamma_2 \\ \vdots \\ \gamma_m \end{pmatrix} = \begin{pmatrix} b_1 \\ b_2 \\ \vdots \\ b_m \end{pmatrix} \tag{2}$$

If the Kutta condition is $\gamma_1 = 0$, then it is clear that the first column of the matrix $[A_{ij}]$ is redundant, hence the system can be reduced to the following:

$$\Longrightarrow \quad \begin{pmatrix} a_{12} & a_{13} & \cdots & a_{1m} \\ a_{22} & a_{23} & \cdots & a_{2m} \\ \vdots & \vdots & \ddots & \vdots \\ a_{m2} & a_{m3} & \cdots & a_{mm} \end{pmatrix} \begin{pmatrix} \gamma_2 \\ \gamma_3 \\ \vdots \\ \gamma_m \end{pmatrix} = \begin{pmatrix} b_1 \\ b_2 \\ \vdots \\ b_m \end{pmatrix} \tag{3}$$

So now we have m equations with $m-1$ unknowns. The system can then be solved by SVD. Since the size of the system is slightly smaller, the CPU time required will also be slightly shorter.

When more than one aerofoil are involved, the Kutta condition will need to be imposed on each of the aerofoils. There will be a few more extra equations, but the same process can be followed and no extra numerical difficulty should result. Lift (or, more correctly, downward force) can be calculated by taking the total circulation around the aerofoils and using the Kutta-Joukowski theorem: $L = -\rho U_\infty \Gamma$. Alternatively the surface pressure could be integrated to give the lift force. This will, of course, give the same result.

4 The Formula-I Multi-Aerofoil Spoiler

The original design is a high downward force spoiler used at circuits where the emphasis is on high corner speed rather than straight line speed. The spoiler is basically a slotted wing which can be modelled as a three-aerofoil system. Unlike an inverted aircraft wing with flaps, in which the flaps are small compared with the fore aerofoil, the 3 aerofoils in the spoiler are of comparable sizes and the middle and rear aerofoils are at large angles to the freestream direction (fig.3). This configuration offers the following advantages:

a. *If the aerofoils do not stall*, they will provide a strong downward force.

b. The centre of pressure is kept as rearward as possible, while the spoiler would situate within a given race specification of 70 cm behind the centre of the rear axle.

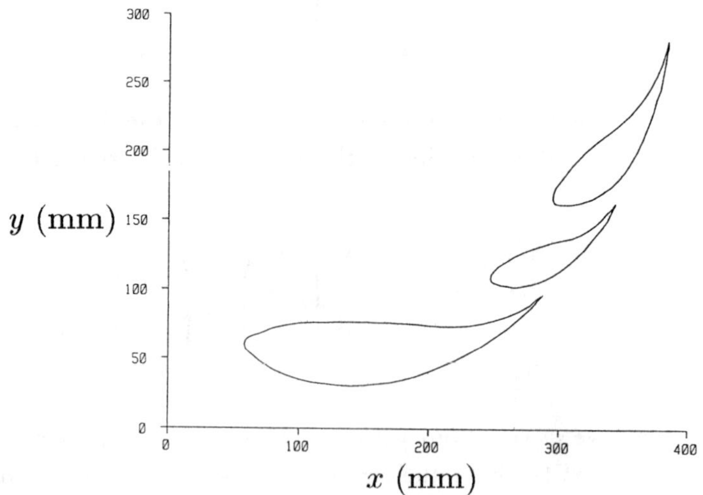

Fig.3 Geometry of the original 3-foil spoiler design.

With the original design (fig. 3), problems of downward force loss and increased drag forces were being experienced at higher angles of attack. The situation was worst when the car has to slow down to turn round corners.

An initial calculation using approximately 50 elements per aerofoil shows that the adverse pressure gradient on the suction side close to the nose is intense (fig.4). It is highly likely that separation occurs close to the leading edge at all speeds. But at high speeds (~ 200 mph) the separated boundary layer may reattach downstream of separation, forming a separation bubble. This would reduce the downward force loss. The flow field shown in fig.5 suggests that the problem is probably due to the sharpness of the nose. The adverse pressure gradient on the middle and rear aerofoils were also inspected, which was found to be moderate such that if the fore aerofoil does not stall, then attached flow on the other aerofoils could be guaranteed.

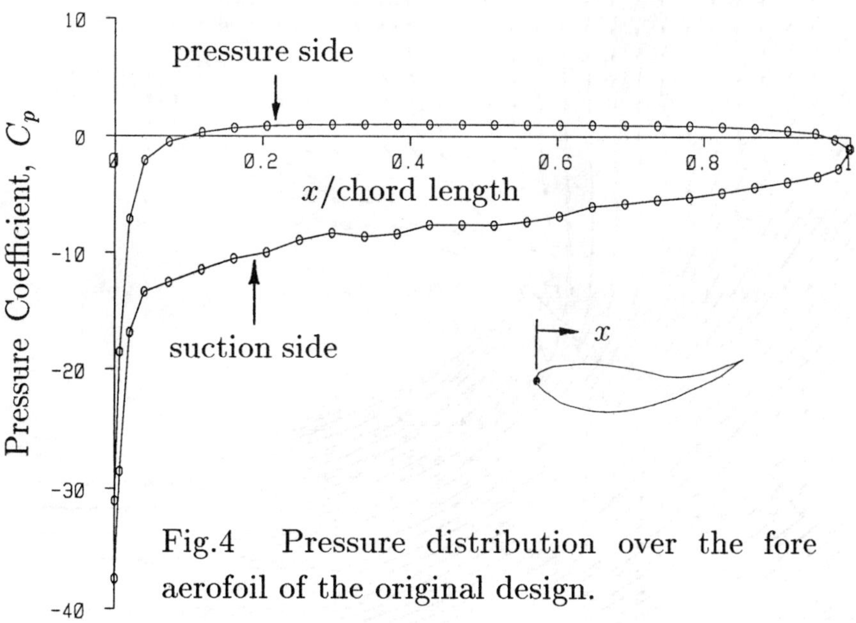

Fig.4 Pressure distribution over the fore aerofoil of the original design.

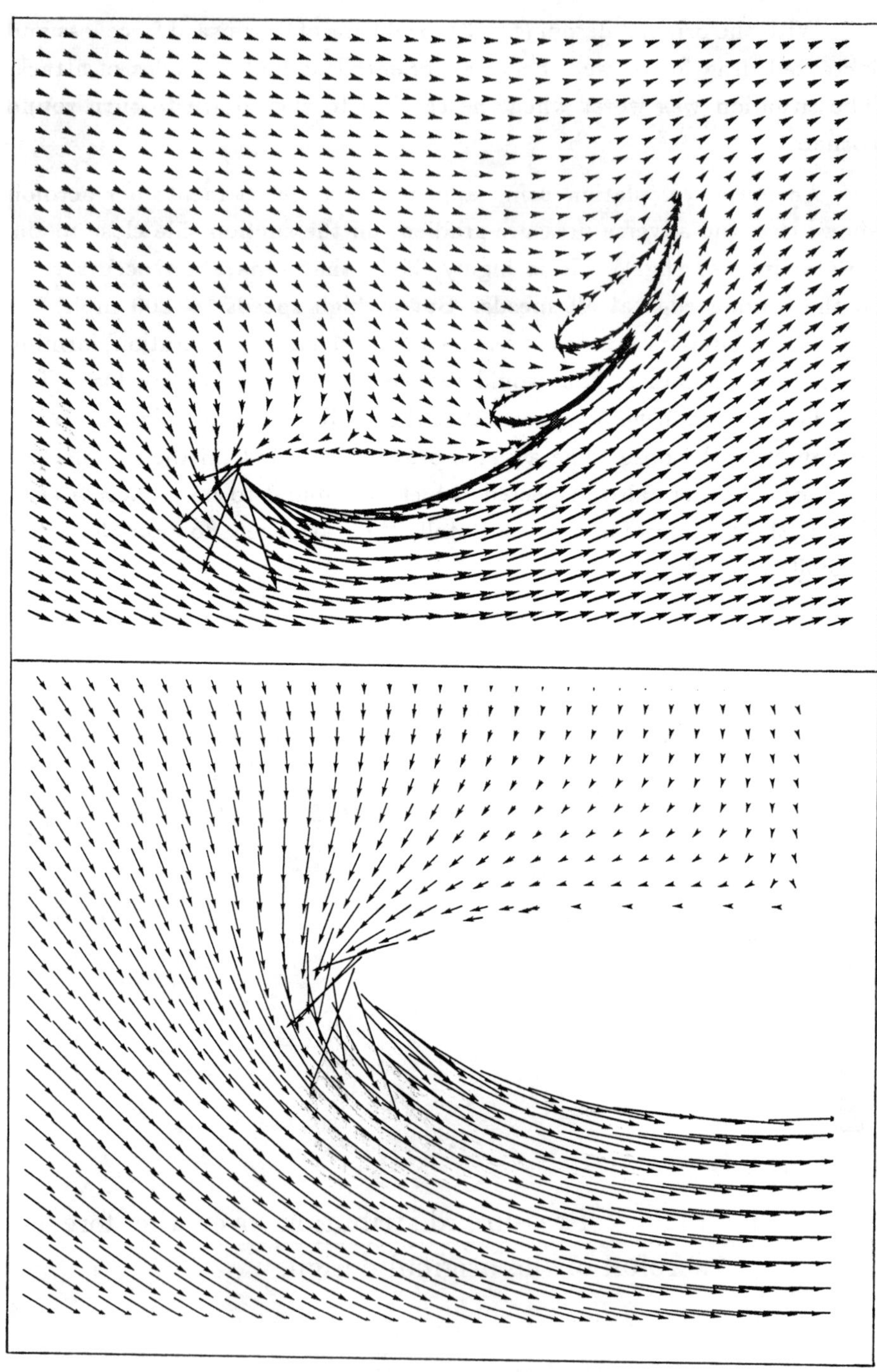

Fig.5 Potential flow field over the original spoiler design.

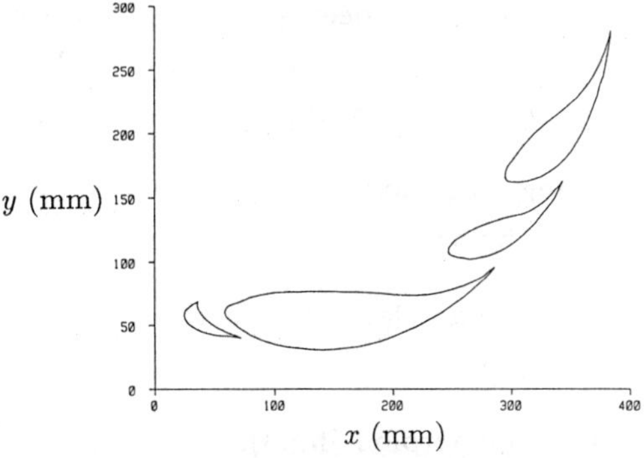

Fig.6 Geometry of modified design with leading edge slat: Position 1.

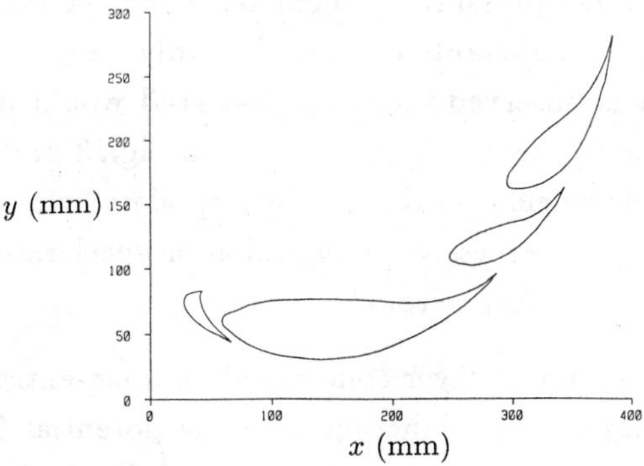

Fig.7 Geometry of modified design with leading edge slat: Position 2.

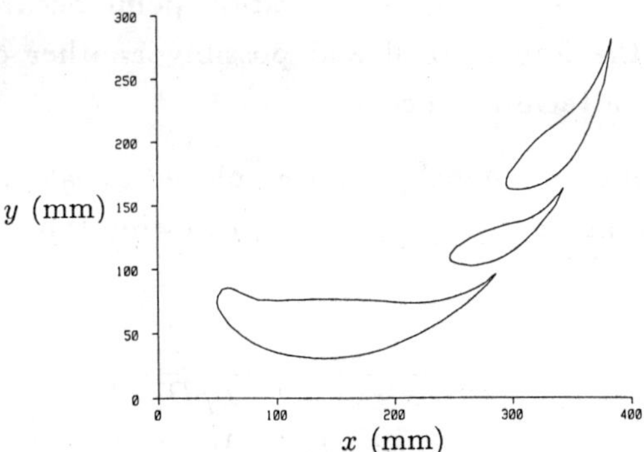

Fig.8 Geometry of modified design with raised nose.

Numerical experiments have been performed on a number of modifications aiming at reducing the adverse pressure gradient whilst keeping alterations to a minimum. Three examples are given here:

a. Adding of a leading edge slat close to the leading edge of the fore aerofoil: position 1 (fig.6);

b. Adding of a leading edge slat close to the leading edge of the fore aerofoil: position 2 (fig.7);

c. Raised nose of the fore aerofoil (fig.8).

The results on the pressure distribution are respectively shown in fig.9, fig.10 and fig.11. It is obvious from these graphs that the suction peak and the adverse pressure gradient occurred on the suction side of the fore aerofoil are substantial reduced. Only very moderate adverse pressure gradient is observed implying that stall would be unlikely, even at relatively lower speeds ($\sim$ 50 mph). Fig.12, fig.13 and fig.14 show the two-dimensional flow field over the modified spoilers, which show that with these modifications, no excessive acceleration or deceleration occurs round the leading edge of the fore aerofoil.

A turbulent boundary layer code based on a lag-entrainment method [4] was also employed in conjunction with the potential flow calculation. The separation points, if any, were computed for each geometry. Reattachment could occur, but is difficult to predict. The modifications considered here all produced encouraging, non-separating results. The original design gave an expected separation point occuring around the leading edge of the fore aerofoil and possibly another one towards its trailing edge if re-attachment occurs.

A simple lifting-line model [3] was employed to calculate the induced drag (drag-due-to-lift) on the spoiler. The following table summarises the results:

Design	c_l	C_{Di} ($\times 10^{-2}$)	L/D_i ratio	% Change in Pitching Moment
Original	2.23	9.36	14.05	-
Slat 1	2.32	9.72	14.06	+0.7%
Slat 2	2.30	9.40	14.42	−1.6%
Raised Nose	2.27	9.58	13.94	+1.1%

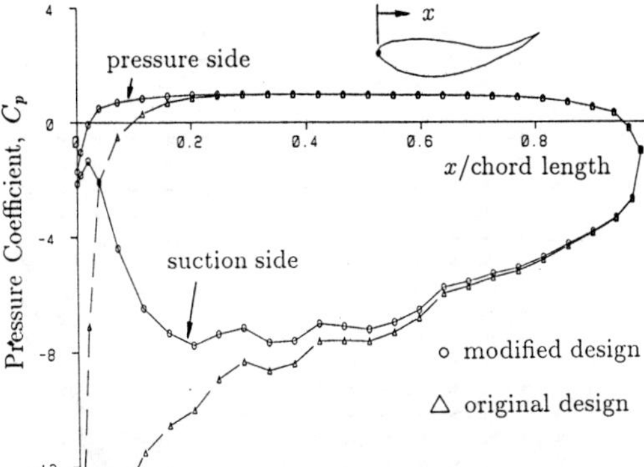

Fig.9 Pressure distribution over the fore aerofoil of the modified design with leading edge slat: Position 1.

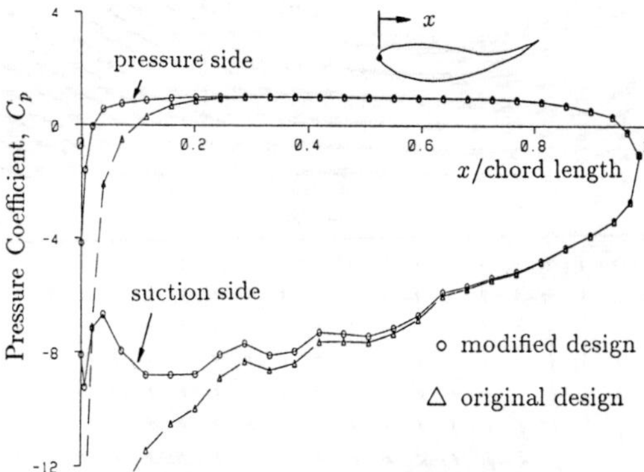

Fig.10 Pressure distribution over the fore aerofoil of the modified design with leading edge slat: Position 2.

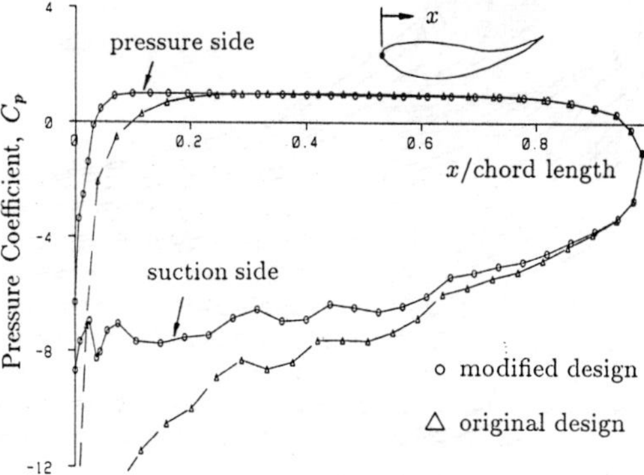

Fig.11 Pressure distribution over the fore aerofoil of the modified design with raised nose.

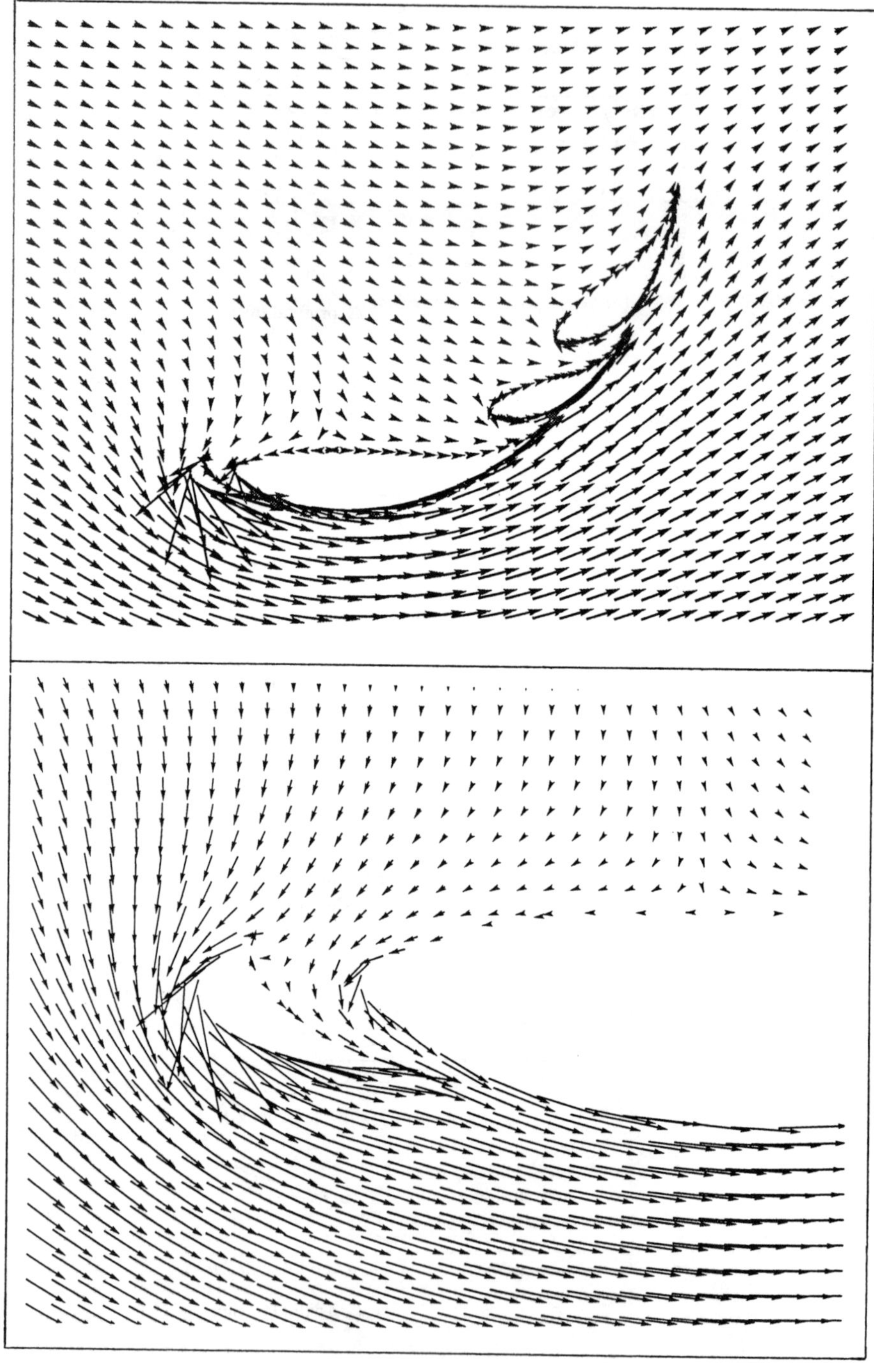

Fig.12 Potential flow field over the modified design with leading edge slat: Position 1.

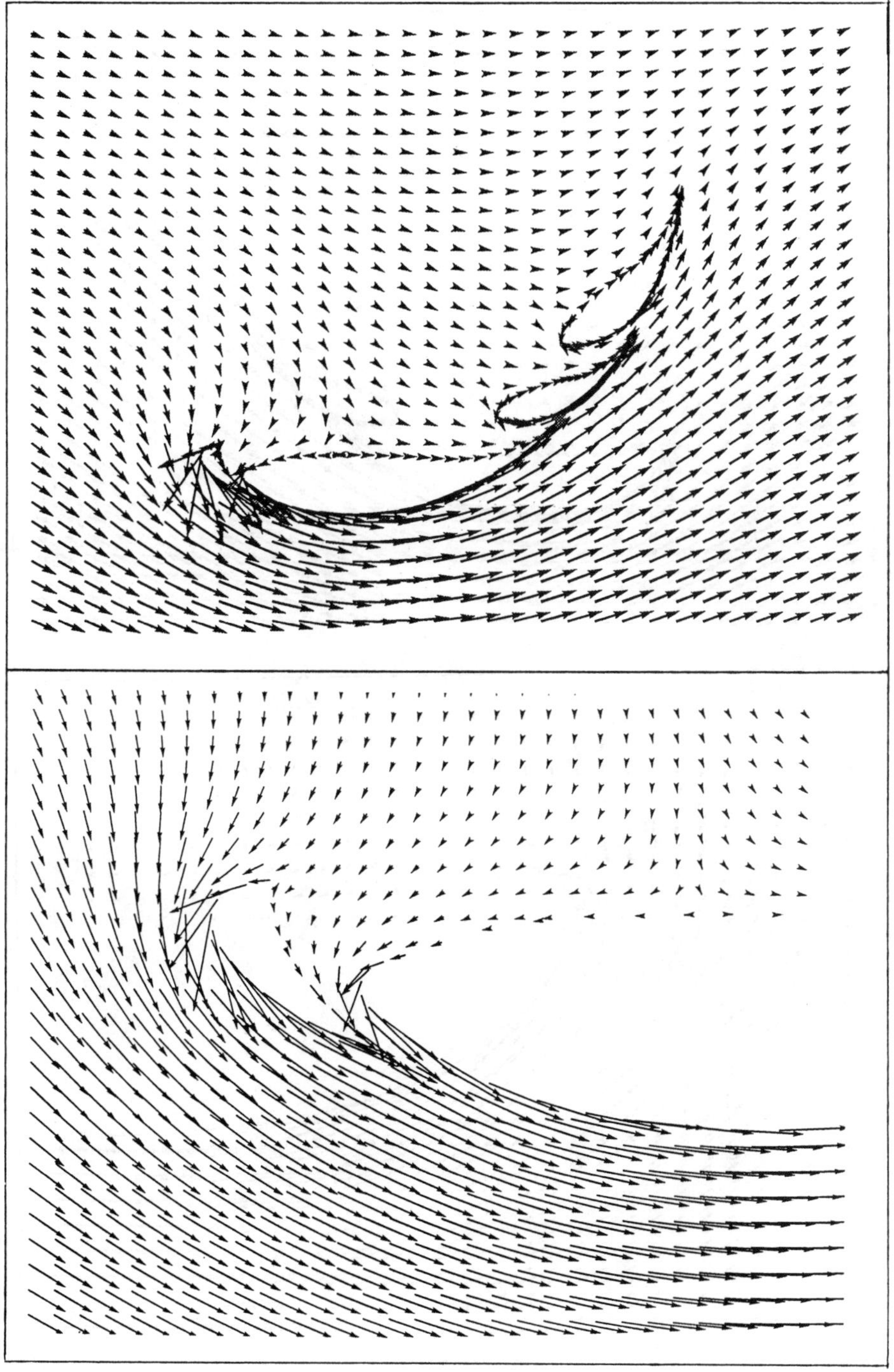

Fig.13 Potential flow field over the modified design with leading edge slat: Position 2.

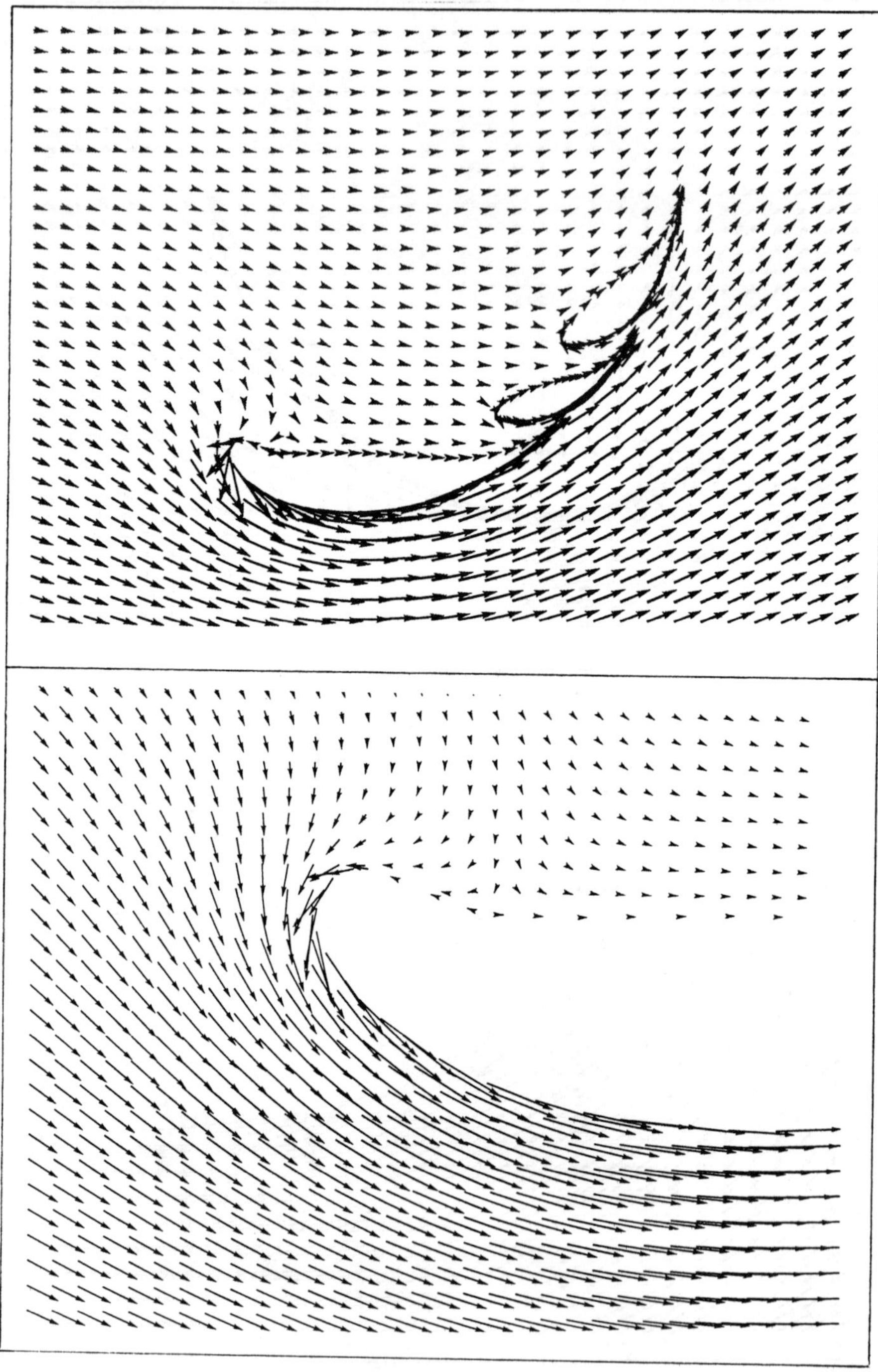

Fig.14 Potential flow field over the modified design with raised nose.

where c_l = Local (2-D) lift coefficient, downward direction; C_{Di} = Total (3-D) induced drag coefficient, L/D_i = Total Lift-to-Induced-Drag ratio, and the pitching moment here is taken about the centre of pressure.

In terms of c_l and L/D_i the three alternative designs have very similar performance, while the raised noise option provides an extra pitching moment of 1.1% . This improvement is enhanced by up to 3.5% when the pitching moment is taken about the rear wheel axle. Taking into account the ease of modification, the raised nose option seems to be preferrable.

5 Conclusions

Of the alterations tested the optimum choice was the large radius raised nose (fig.8). Although the leading edge slats gave a more desirable pressure distributions (i.e. lower adverse pressure gradient) around the fore aerofoil, the slats moved the centre of pressure forward by about 2.5 cm. This gave an approximate reduction of 3.5% in pitching moment around the rear wheel axle.

The magnitude of lift coefficients did not vary considerably, as most of the alterations were directed towards the reduction of adverse pressure gradient and the risk of flow separation which would eventually lead to higher spoiler effectiveness at lower speeds. Since the present calculation is a potential flow approximation, the results are qualitative and the real improvement percentages are not available until full scale experimental analysis has been performed. However, it appears from the results that the large radius raised nose is the best solution of the twenty or so designs tested, and it is also easy to implement. Thus, this was the recommended design for wind tunnel testing this racing season.

This clearly demonstrates the ease of use and economy of CVBEM in early stages of aerodynamic design. Huge savings on wind tunnel time (which could amount up to £1000 per day) and materials have been made. The project is now in the experimental stage with an intention to evolve the process into three-dimensions and encompass the flow over the entire car.

6 References

1. Chiu, T.W., "A two-dimensional seconder-order vortex panel method for the flow in a cross-wind over a train and other two-dimensional bluff bodies", *J. Wind Eng. Ind. Aerodyn., 37 (1991) 43-64.*

2. Press, W.H., Flannery, B.P., Teukolsky, S.A. and Vetterling, W.T., "Numerical recipes (FORTRAN version)", *Cambridge University Press, 1989.*

3. Anderson, J.D., "Fundamentals of Aerodynamics", *McGraw-Hill, Inc., 1984.*

4. Green, J.E., Weeks, D.J. and Brooman, J.W.F., "Prediction of turbulent boundary layers and wakes in compressible flow by a lag-entrainment method", *ARC Reports and Memoranda 3791, 1973 (Royal Aerospace Establishment, Farnborough, Hants.).*

SECTION 3: VISCOUS FLOW AND TURBULENCE MODELS

Description of Viscous-Inviscid Interaction Using Boundary Elements

H. Schmitt

Institute of Theoretical Fluid Mechanics, Deutsche Forschungsanstalt für Luft- und Raumfahrt (DLR), Bunsenstr. 10, D-3400 Göttingen, Germany

ABSTRACT

Steady two- and three-dimensional flows of a viscous incompressible fluid about an arbitrary body are considered. Far from the body, the fluid can be considered as inviscid and the flow can be calculated with use of boundary element methods. In the neighborhood of the body and in its wake, the viscosity has to be taken into account, and the flow can be calculated by means of finite difference methods for example. If the Reynolds number is high, the regions, in which the viscosity is essential, that are the boundary layers at the surface, dead air regions, and free shear layers, are thin.

The interactions between these viscous and inviscid regions of the whole flow field have been treated by different methods applied to the respective regions and by different assumptions concerning the interaction effects. This will be discussed by means of three examples.

As first example, the two-dimensional flow around an airfoil is considered, in which the boundary layer separates from the airfoil and an adjacent dead air region is formed. As second example, the flow around a prolate spheroid at incidence is treated, in which case separation leads to a thin free vortex layer. A third example is given by a turbulent circular jet exhausting normally from a flat plate into a stream flowing parallel to the plate. The viscous jet is bent by the uniform inviscid flow and the inviscid flow is modified in the neighborhood of the jet.

1. BASIC EQUATIONS

The three-dimensional steady flow of a viscous incompressible fluid about an arbitrary body is considered. If the flow is laminar, the nondimensional balance equations for mass and momentum are:

$$div\ \mathbf{v} = 0\ , \tag{1}$$

$$\mathbf{v} \cdot grad\ \mathbf{v} = -\ grad\ p + \frac{1}{Re}\ \Delta\ \mathbf{v}\ . \tag{2}$$

Here, $\mathbf{v}$ is the relation of the velocity to its absolute value V_∞ in the uniform flow at infinity. The quantity p is the deviation of the pressure from the value p_∞ at infinity related to ρV_∞^2, where ρ is the density. The position vector $\mathbf{x}$ is related to a characteristic length a. $Re = aV_\infty/\nu$ is the Reynolds number of the uniform flow, where ν is the kinematic viscosity.

If the flow is turbulent, the continuity equation (1) holds for the mean flow. Instead of the Navier-Stokes equation (2) the Reynolds equation can be used to describe this flow (see for example Rotta [34] or Hinze [13]).

The boundary conditions are:

$$\mathbf{v} = \mathbf{v}_S \qquad on\ the\ body\ surface\ , \tag{3}$$

$$\lim_{|\mathbf{x}| \to \infty} \mathbf{v} = \mathbf{i}\ . \tag{4}$$

Here, $\mathbf{i}$ is a unity vector in the direction of the flow at infinity. The velocity $\mathbf{v}_S$ on the body surface is given by suction or blowing through it.

In the following, only flows with high Reynolds numbers are considered; then mathematically, the problem is a singular perturbation problem (see for instance Kevorkian and Cole [20]). The zeroth order outer solution is that of the inviscid problem. For this, when the velocity is irrotational, the disturbance velocity potential φ is introduced:

$$\mathbf{v} = \mathbf{i} + grad\ \varphi\ . \tag{5}$$

With this, the mass-conservation equation (1) becomes:

$$\Delta\ \varphi = 0\ . \tag{6}$$

Since the differential equation (6) is elliptic, a modification in a small region modifies the whole inviscid flow field. This occurs, if the flow separates from the surface of the body and forms a wake; then, the fluid in the inviscid region is split by this wake and forced to go an other way. This influence of the viscous effects on the inviscid flow field is taken into account by boundary conditions at the surface of the wake. In the limit case $Re \to \infty$, not only the thickness of the boundary at the body surface becomes zero, but also the wake region degenerates into a vortex sheet originating from the body or forms a dead air region of finite thickness and length. The vortex sheet and the boundary of the dead air region are stream surfaces, across which the pressure is continuous. Therefore, the boundary conditions for the inviscid problem are:

$$\mathbf{v} \cdot \mathbf{n} = \mathbf{v}_S \cdot \mathbf{n} \qquad on\ the\ surface\ of\ the\ body\ , \tag{7}$$

$$\mathbf{v} \cdot \mathbf{n} = 0 \qquad on\ the\ vortex\ sheet\ and\ on\ the\ boundary\ of\ the\ dead\ air\ region\ , \tag{8}$$

$$p_+ = p_- \qquad across\ the\ vortex\ sheet\ and\ across\ the\ boundary\ of\ the\ dead\ air\ region\ . \tag{9}$$

Here, $\mathbf{n}$ is the unity vector in the direction of the outward normal, and the two sides of the vortex sheet are denoted by + and − , respectively.

The potential flow can be calculated by means of superposition of fundamental solutions of Laplace's equation (6). In the following, the potentials $\varphi^*(\mathbf{x}, \mathbf{q})$ of sources and the potentials $\partial\varphi^*/\partial n$ of source doublets are used, which are located on the surface S consisting of the surfaces of the body, of the vortex sheet, and of the boundary of the dead air region:

$$\varphi(\mathbf{x}) = -\int_S \sigma(\mathbf{q})\, \varphi^*(\mathbf{x}, \mathbf{q})\, dS(\mathbf{q}) + \int_S \mu(\mathbf{q}) \frac{\partial \varphi^*(\mathbf{x}, \mathbf{q})}{\partial n(\mathbf{x})}\, dS(\mathbf{q}) \, , \tag{10}$$

$$\varphi^*(\mathbf{x}, \mathbf{q}) = \begin{cases} \dfrac{1}{r(\mathbf{x}, \mathbf{q})} & \text{for three} - \text{dimensional flow,} \\ \ln \dfrac{1}{r(\mathbf{x}, \mathbf{q})} & \text{for two} - \text{dimensional flow.} \end{cases} \tag{11}$$

The strength σ of the sources and the strength μ of the source doublets depend on the position $\mathbf{q}$ on the surface, $r(\mathbf{x}, \mathbf{q})$ is the distance between the reference point $\mathbf{x}$ and the source point $\mathbf{q}$, $\partial/\partial n$ is the derivative in the direction of the source doublet, which is directed perpendicular to the surface S.

The strengths of the sources and the source doublets are determined by putting (10) into the boundary conditions (7) and (8). In the special case, in which only sources are used, the following Fredholm integral equation of the second kind results for a point $\mathbf{p}$ on the surface S:

$$m\,\pi\,\sigma(\mathbf{p}) - \int_S \frac{\partial \varphi^*(\mathbf{p}, \mathbf{q})}{\partial n(\mathbf{p})}\, \sigma(\mathbf{q})\, dS(\mathbf{q}) = \mathbf{n}(\mathbf{p}) \cdot (\mathbf{v}_S - \mathbf{i}) \, , \tag{12}$$

where $m = 1$ for two-dimensional flow and $m = 2$ for three-dimensional flow (see also Brebbia, Telles, Wrobel [5] and Jaswon [19]). For the numerical solution of equation (12), boundary elements on the surface of the body, of the vortex sheet, and of the dead air region are introduced.

For finite but high Reynolds numbers, all viscous regions are finite but thin layers; these are boundary layers adjacent to a body surface, free shear layers separating from a body surface, and free shear layers at the boundary of a dead air region. For these layers, the equations (1) and (2) reduce to Prandtl's boundary layer equations, which describe the first order inner solution of the singular perturbation problem.

For a three-dimensional laminar boundary layer, these equations are given in [42] for example in orthogonal curvilinear coordinates. For turbulent two-dimensional wall shear flows and free shear flows, the boundary layer equations are given for example by Rotta [34] or Hinze [13] in Cartesian coordinates. Hinze also considers turbulent three-dimensional free shear flows. Turbulent three-dimensional boundary layers are considered by Patel and Choi [29] using cylindrical polar coordinates, Tanaka [44] using orthogonal curvilinear coordinates, and Barberis [1] and Cebeci et al. [6] using nonorthogonal curvilinear coordinates.

The treatment of the viscous-inviscid interactions outlined so far can be used as an iterative procedure. That is, the approximate shape of the wake assumed for the calculation of the inviscid flow is improved by the results for the viscous flow.

The method can be extended to the flow around several bodies. In this case, the thicknesses of the boundary layers at the surfaces of every two bodies must be small in comparison with the thickness of the inviscid flow region between them. Otherwise, the whole flow in this region must be calculated with help of equations (1) and (2). An example of such a flow has been given in [40], where a prolate spheroid moving near ground has been considered.

In the following, three examples will be discussed, for which the interaction between the viscous and inviscid regions of the flow field play an important part.

2. FLOW AROUND AN AIRFOIL

2.1. Attached flow

As a first example, the two-dimensional flow around an airfoil with a sharp trailing edge at an angle of attack is considered. In this case, the influence of viscosity on the inviscid flow can be demonstrated. That is, the calculation of the inviscid problem without the boundary conditions (8) and (9) on the surface of a wake leads to streamlines, which are not shown by a real fluid. A model of this hypothetical inviscid flow is the Hele-Shaw flow around the airfoil (see for instance [2]) shown in figure 1. The fluid from the lower side flows around the sharp trailing edge, with very high velocity at this edge. On the upper side, the fluid separates from the airfoil before reaching the trailing edge. Thus, the Kutta-Joukowski condition is violated, which demands finite velocities at the trailing edge (see [2]).

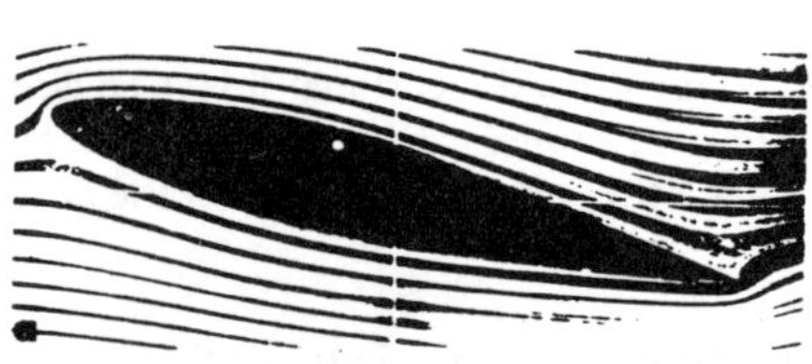

Fig. 1. Hele-Shaw flow past an inclined airfoil (Werlé [45])

Fig. 2. High Reynolds number flow past an inclined airfoil measured by Prandtl [2] (from right to left)

The inviscid flow of a real fluid only can be calculated by taking into account the existence of the wake. For high Reynolds numbers, the streamlines are given in figure 2. The viscosity of the fluid modifies the inviscid flow in the region near the trailing edge in such a way, that the Kutta-Joukowski condion is fulfilled (see also [2]) and the separation occurs at the trailing edge. Before this interpretation had been found, the condition was known as an empirical rule.

2.2. Separated flow

If the angle of attack is increased, the separation point moves to the upper side of the airfoil, and a dead air region is formed between the separation point and the trailing edge. Figure 3 shows streamlines of this flow, where the separating boundary layer is laminar. Jacob [14, 15, 18] has given a method to calculate this flow, which shall be explained in the following.

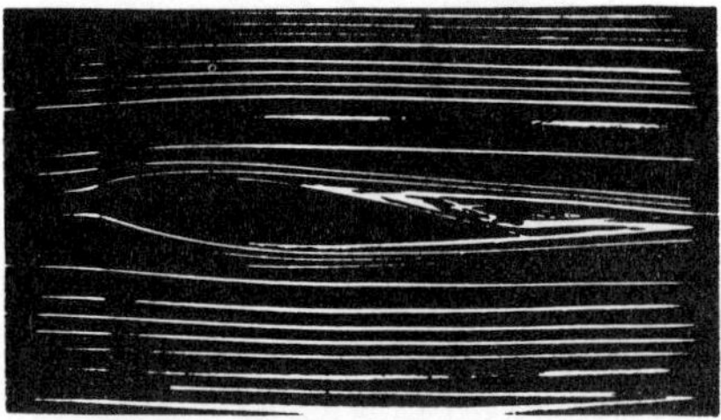

Fig. 3. Separation on an inclined airfoil (Werlé [45])

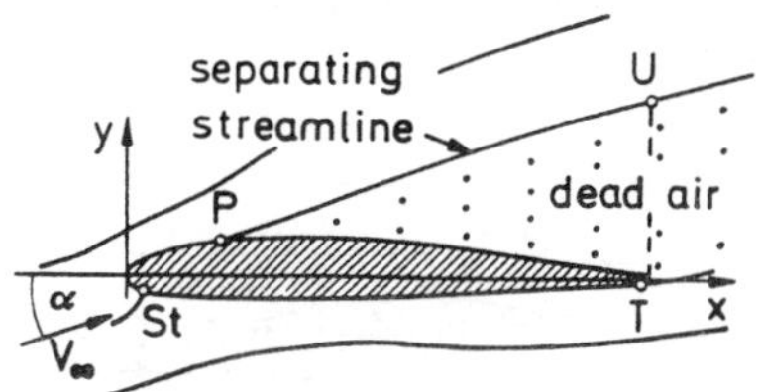

Fig. 4. Flow with simulated dead air (Jacob [15, 18])

In the two-dimensional case, a stream function ψ can be introduced:

$$\mathbf{v} = \mathbf{i} + \mathit{grad}\,\varphi + \mathit{rot}\,(\psi\,\hat{\mathbf{z}}) \ . \tag{13}$$

Here, $\hat{\mathbf{z}}$ is a unity vector, which is perpendicular to the flow. For irrotational flow, the stream function ψ also satisfies the Laplace equation (6), and it can be solved by superposition of fundamental solutions:

$$\psi(\mathbf{x}) = \frac{1}{2\pi} \int_S v_t(\mathbf{q})\, \psi^*(\mathbf{x}, \mathbf{q})\, dS(\mathbf{q}) \ , \tag{14}$$

$$\psi^*(\mathbf{x}, \mathbf{q}) = \ln \frac{1}{r(\mathbf{x}, \mathbf{q})} \ . \tag{15}$$

Here, $v_t = \mathbf{t} \cdot \mathbf{v}$, where $\mathbf{t}$ is a unity vector in the direction of the tangent to the boundary S.

Then with use of (10) without source doublets for φ, the following integral equation holds (see Martensen and v. Sengbusch [24]):

$$\begin{aligned} v_t(\mathbf{p}) + \frac{1}{\pi} \frac{\partial}{\partial n(\mathbf{p})} \int_S v_t(\mathbf{q})\, \psi^*(\mathbf{p}, \mathbf{q})\, dS(\mathbf{q}) &= \\ = 2\, \mathbf{t}(\mathbf{p}) \cdot \mathbf{i} - \frac{1}{\pi} \frac{\partial}{\partial t(\mathbf{p})} \int_S v_n(\mathbf{q})\, \varphi^*(\mathbf{p}, \mathbf{q})\, dS(\mathbf{q}) \ . & \end{aligned} \tag{16}$$

Here, $v_n/2\pi$ with $v_n = \mathbf{n} \cdot \mathbf{v}$ is the source density σ, and v_t is a distribution of vortices.

Jacob has used this integral equation neglecting the sources v_n for the calculation of the flow with separated dead air region as shown in figure 4. His basic idea is the simulation of the dead air region by an additional inviscid flow issuing from the airfoil, which displaces the original inviscid flow from the dead air region. This additional flow is produced by a source distribution on the surface of the airfoil between the points P and T. The total strength of this sources is determined by additional conditions. The differences of the velocities in two points of the dead air region are small in comparison with the respective differences in the inviscid flow region, because the fluid material in the dead air originates from the surface of the airfoil, and because also reverse flows occur in that region. As a consequence, the pressure differences are small; this is confirmed by experimental data (see for example Krämer [21]). In order to simulate a constant pressure in the wake, Jacob assumes equal pressure in three points of the dead air region, namely in the two separation points P and T on the upper and the lower surfaces and in the point U, which is situated on the upper separating streamline above the trailing edge.

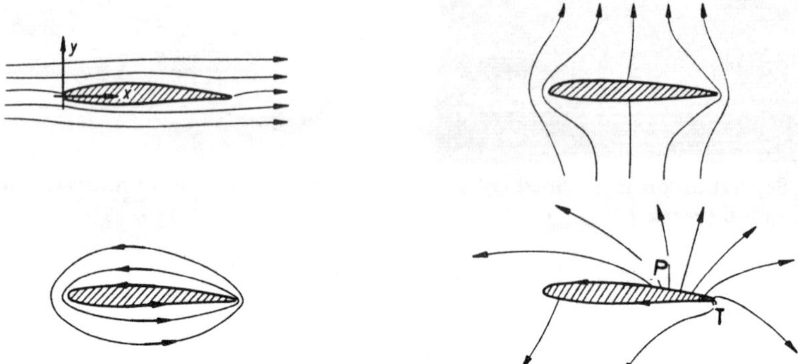

Fig. 5. Four elementary flows (Jacob [15])

The flow of figure 4 is assumed to be a superposition of four elementary flows shown in figure 5:

$$\mathbf{v} = \mathbf{v}^{(1)} \cos\alpha + \mathbf{v}^{(2)} \sin\alpha + b\mathbf{v}^{(3)} + d\mathbf{v}^{(4)} . \tag{17}$$

Here, $\mathbf{v}^{(1)}$ and $\mathbf{v}^{(2)}$ are uniform flows in the x- and y-direction without circulation around the airfoil and without sources. $\mathbf{v}^{(3)}$ is a circulatory flow around the airfoil with vanishing velocity at infinity and without sources. $\mathbf{v}^{(4)}$ is an outflow produced by a given source distribution on the section PT of the surface of the airfoil. The constants b and d are determined by the two conditions, that the pressure in the three points P, T, and U has to be the same.

The calculated inviscid flow including the inviscid model of the dead air flow assumes given separation points P and T on the upper and the lower surface of the airfoil. With help of this inviscid flow, the boundary layer on the upper side of the airfoil is calculated and a new separation point $P' \neq P$ is determined. This viscous-inviscid interaction problem is solved by an iterative procedure; the angle α of attack is varied, until P' tends to P. Thus compatible inviscid and viscous solutions are found.

For the calculation of the boundary layer, the integral method of Rotta [32, 33] has been used, who considered two ordinary differential equations for the momentum thickness δ_2 and the energy thickness δ_3 of the boundary layer. The first results from integration of the momentum equation from the wall to a value lying outside the boundary layer (see for example Schlichting [35, p.676]):

$$\frac{d\delta_2}{ds} + \delta_2 \frac{H_{12}+2}{u_e} \frac{du_e}{ds} = \frac{1}{2} c_f . \tag{18}$$

The second equation is obtained by multiplying the equation of motion by the velocity and then integrating from the wall to a value lying outside the boundary layer:

$$\frac{d\delta_3}{ds} + \delta_3 \frac{3}{u_e} \frac{du_e}{ds} = c_D . \tag{19}$$

Here, s is the distance measured along the surface, u_e is the velocity at the edge of the boundary layer, and $H_{12} = \delta_1/\delta_2$ is the shape factor formed with the displacement thickness δ_1; c_f is the skin friction coefficient, and c_D is the dissipation coefficient.

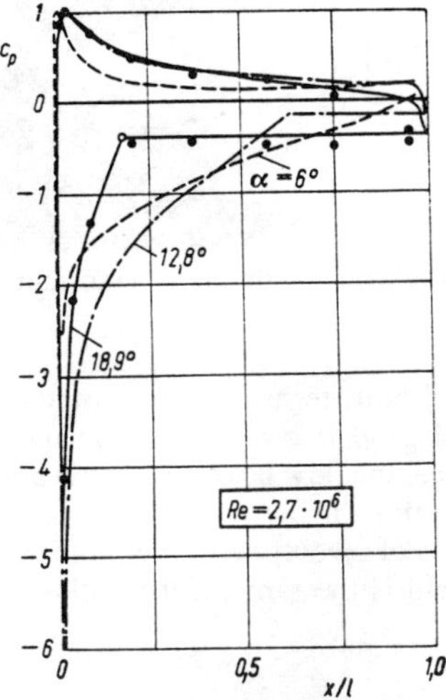

Fig. 6. Pressure distribution of the airfoil; theory (Jacob [15]):
– – – no separation, $\alpha = 6°$,
– · – separation at $x/l = 0.629$, $\alpha = 12.75°$,
—— separation at $x/l = 0.179$, $\alpha = 18.87°$;
experiments from [22]:
• • • separation at $x/l \approx 0.18$, $\alpha = 17.9°$.

Figure 6 shows results for the pressure coefficient

$$c_p = 2p = 1 - v^2 \tag{20}$$

of the upper and the lower side of the airfoil as a function of x/l, where l is the chord length. It can be seen, that the point of separation moves to the front of the airfoil, if the angle of attack increases. For the case $\alpha = 18.9°$, in which the dead air region has its largest extension, Jacob has found very good agreement with experimental results [30]. The calculated boundaries of the dead air region have been given in figure 4.

The method has been extended to the flow around multi-element airfoil systems with separation (Jacob [18]) and around airfoils and airfoil systems moving near ground and situated in a wind-tunnel (Steinbach [43, 18]). For the treatment of three-dimensional flows, that is for flows around wings, Jacob has combined his two-dimensional theory with an inviscid three-dimensional lifting surface theory [16]. In [17], he has given improvements for this method and an extension to include ground effects.

3. FLOW AROUND A PROLATE SPHEROID

In the following, the flow around a three-dimensional body, a wing for example, will be discussed. If the wing is large, that is if it has a high aspect ratio, for most of its airfoil sections figure 3 gives the shape of the closed region of separated two-dimensional flow; only in the neighbourhood of the tips, the separated flow has an other shape. On the other side, if the aspect ratio is extremely small, the separation effects of the tips are predominant. An example is a slender prolate spheroid. The flow pattern on the surface for a ratio of the axes of 4.3 visualized using dye is shown in figure 7. The flow forms two open separation lines on both sides of the spheroid and the boundary layers roll up into two longitudinal vortices along these lines. Contrary to the two-dimensional case, no large closed separation region exists.

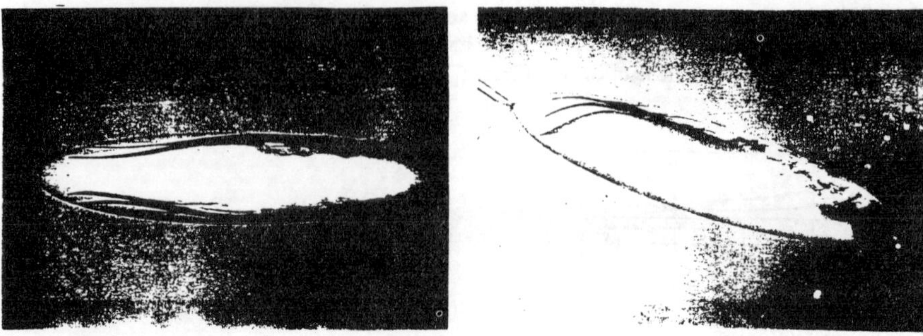

Fig. 7. Prolate spheroid: flow pattern at $\alpha = 20°$, (a) top view, (b) side view (Han and Patel [11])

If the viscosity is neglected completely, that is, without regard of the boundary conditions (8) and (9), the inviscid flow around the spheroid is given by the analytic solution of Heine (see Lamb [22]). If there is no incidence, the flow is axisymmetric and both stagnation points are situated on the axis of revolution. If the angle of incidence increases, both stagnation points move away from the axis of revolution at the same rate of the arc in the plane of symmetry. The whole flow field is the same, if if the flow is reversed.

This problem can also be treated with boundary element methods with use of only sources on the surface, that is, with use of integral equation (12). The comparison of the result with the analytic solution shows, that the approximation by the boundary element solution is better for higher number of elements (see [38]).

On the assumption that the inviscid flow is described by the analytic solution of Heine, the three-dimensional boundary layer equations (see [42]) have been solved for example by Wang [46], Patel et al. [29, 28], Cebeci and Su [7], and Barberis [1]. Geißler [9] has used a boundary element method with a source distribution on the surface. Figure 8 shows the separation line on the surface of the spheroid at an axis ratio 1/6, an angle of incidence of $\alpha = 10°$, and an upstream Reynolds number $Re = 0.8 \cdot 10^6$; here, the large semiaxis of the spheroid is taken as characteristic length.

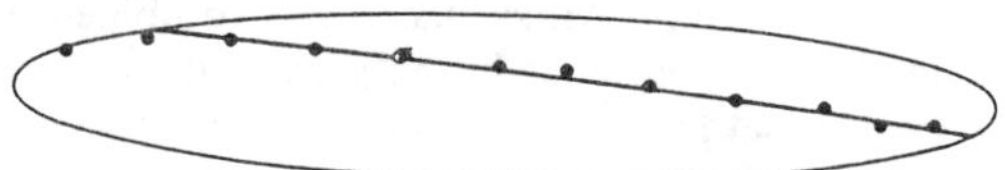

Fig. 8. Circumferential separation line (side view), • experimental, — calculated (Patel and Baek [28])

The inviscid flow solutions discussed so far only describe the displacement effect of the spheroid on the uniform flow. Because of the symmetry for reversal of the flow, no lift acting on the spheroid is produced. In order to improve this solution, the separating viscous layer must be taken into account. Figure 7 suggests, that this can be approximated by the vortex sheet S shown in figure 9 (see [38, 42]). This implies that the rearward stagnation point St_2 is moved to the axis of revolution. An additional argument for this assumption relating to the position of the vortex sheet is the fact, that for a small ratio of the two axes of the spheroid, a linearized theory leads to the position

$z = 0$. Then besides the sources on the surface of the spheroid, source doublets on the vortex sheet have been used (see [38, 42]). In the computation, the effect is taken into account, that according to Ampère's rule a boundary element with a source doublet is equivalent to an element bounded by a vortex filement (see for example [23]).

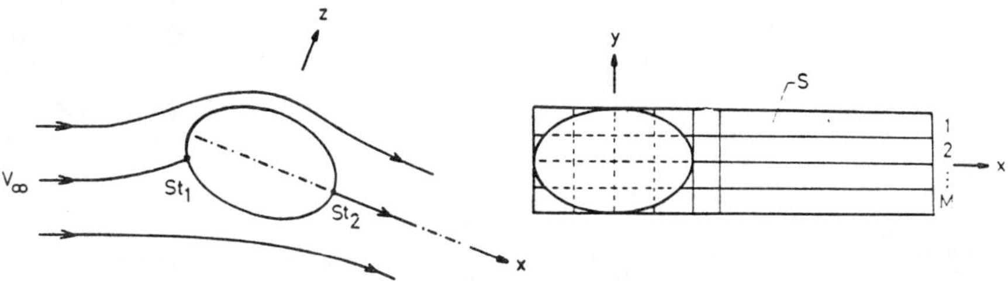

Fig. 9. (a) Qualitative streamlines with regard of a vortex sheet in the case of incidence, (b) shape of the vortex sheet S situated in the plane $z = 0$

Figures 10a and b show, that the consideration of the vortex sheet substantially improves the agreement of the calculated results with those of the measurements.

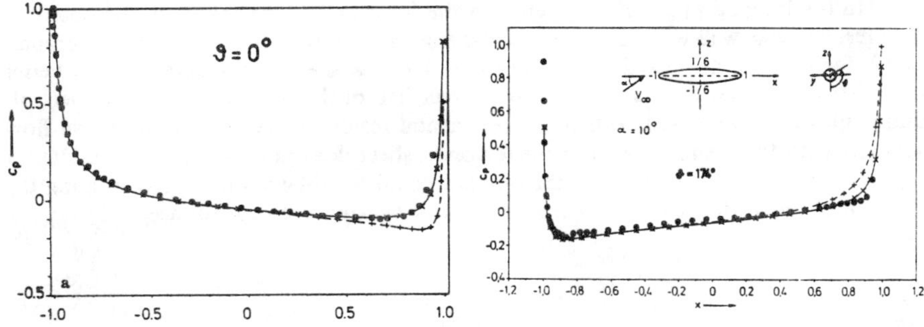

Fig. 10. Pressure coefficient on the surface for $\alpha = 10°$ and an axes ratio of 1/6, calculation with (— x —) and without (+ – +) vortex sheet [39, 38, 42], ⊕⊕⊕ experiments of (a) Böddener [3] and (b) Meier and Kreplin [25]

In figure 11 for a higher angle of attack, a comparison with curves of Panaras and Steger is given, who have used the thin-layer approximation for the Navier-Stokes equations (1, 2), in which only viscous terms in the direction normal to the body are retained. It should be noted, that this method is much more expensive than that discussed here. The results of figure 11 relate to values of ϑ differing by 5°. The influence of this difference is much smaller than that between the different curves. The experimental results do not depend appreciably on the Reynolds numbers, but on the leeward side for $-0.4 \leq x \leq 0.1$; there the lower Reynolds number flow has a higher pressure, because it remains longer laminar.

In the windward plane of symmetry again, the improvement of the calculated values caused by the consideration of the vortex sheet can be seen. The thin-layer approximation is of the same good quality except in the small region $0.9 \leq x < 1$. Further, it can be noted, that for $0.6 \leq x < 1$ the predictions differ from the measurements in opposite directions.

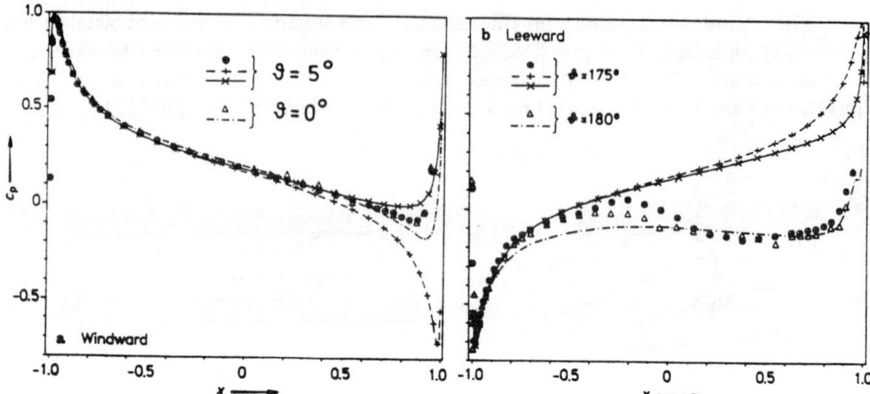

Fig. 11. **Pressure coefficient on the surface for $\alpha = 30°$ and an axes ratio of 1/6,** ⊕⊕⊕ experiments of Meier and Kreplin [25] for $Re = 4.9 \cdot 10^6$, △△△ experiments of Meier [27] for $Re = 6.0 \cdot 10^6$, potential flow calculation [39] with (—×—) and without (+ — +) vortex sheet, — · — thin-layer solution of Panaras and Steger [27] for $Re = 2.2 \cdot 10^7$

In the leeward plane of symmetry on the front half of the spheroid, the potential flow results agree well with the experimental results, as long as the flow remains laminar, that is for $-1 \leq x \leq -0.2$. Particularly for $-0.95 \leq x \leq -0.2$, the agreement is better than that of the thin-layer solution. On the rear half of the spheroid, the thin-layer solution still agrees very well with the experimental results. Contrary, the potential flow solution with the assumption of a plane vortex sheet does not lead to good results for such a high angle of attack. An improvement could be obtained by approximating the position of the vortex sheet in such a way that it became a stream surface.

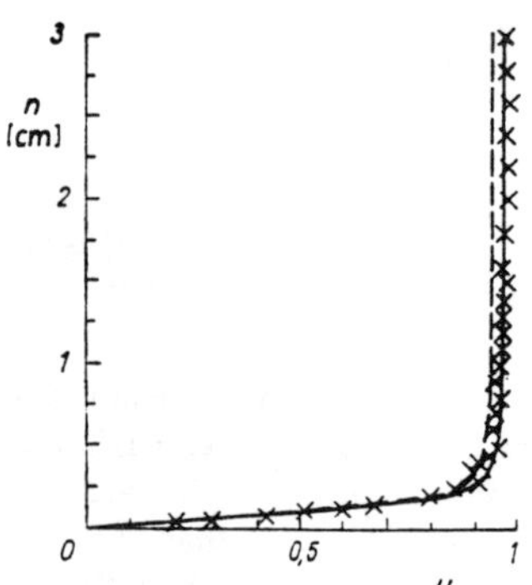

Fig. 12. Velocity u as function of the distance n from the wall for $x = 0.77$, theory with (—) and without (– – –) vortex sheet [41], × × × experiments Coponet [8]

With use of the inviscid flow solution improved by regarding the vortex sheet, a new improved calculation of the boundary layer is possible. This has been accomplished in [41, 42] for the plane of symmetry. Figure 12 shows results for the axes ratio 1/6, the major half axis $a = 1.2\ m$, and the angle of attack of $\alpha = 10°$; again, the essentially better agreement with experimental results can be seen.

4. FLOW AROUND A CROSSING CIRCULAR JET

The geometry of the last example is demonstrated by figure 13; it is the flow along a flat plate, into which a turbulent jet exhausts normally from the plate through a circular orifice. Initially, the jet behaves like a solid body, around which the main fluid flows. On both sides of the quasi-solid jet, vortex sheets of the main stream separate as in the case of the flow around the spheroid at incidence. These vortex sheets merge with the material of the jet, until only one vortex sheet survives. Thus, the initially circular cross-section of the jet is transformed into the cross-section of the rolling up vortex sheet.

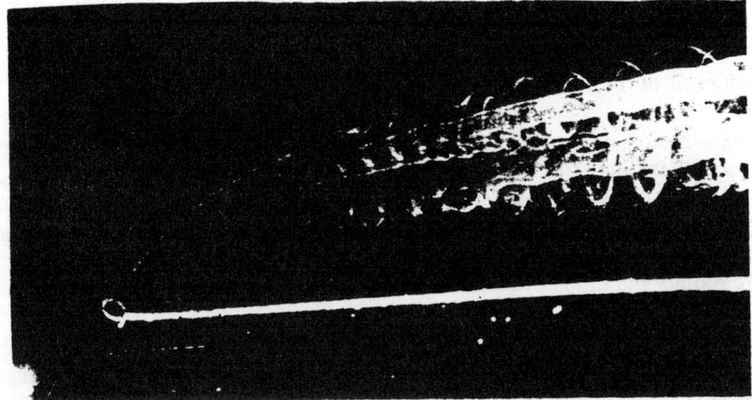

Fig. 13. A low Reynolds number smoke jet in a crosswind (from Hacket and Miller [10])

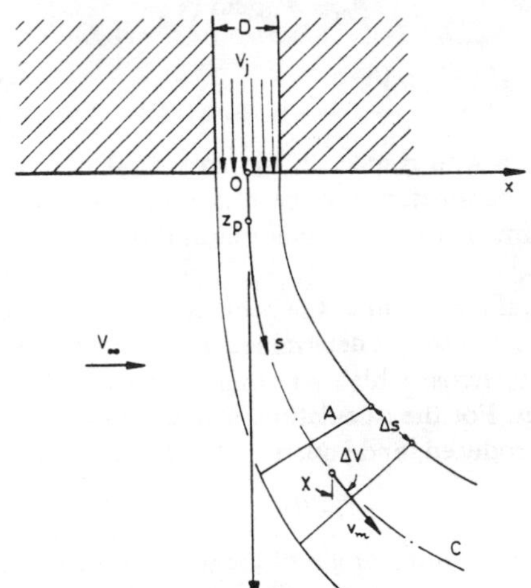

Fig. 14. Circular turbulent jet in a crosswind

The interaction of the jet and the uniform main flow can be treated as follows. In the first step, the main flow is treated as uniform everywhere, except within the boundary of the jet, where it does not exist. Also the boundary layer on the flat plate is disregarded. In the second step, the bent jet flow can be calculated with use of modelled balance equations for mass and momentum of a volume element ΔV as sketched in figure 14. In the limit case $\Delta s \to 0$, where s is the arc length along the centre line C of the jet, these balance equations are given by (see [36, 37]):

$$\rho \frac{d}{ds}(Av_m) = \rho e \ , \tag{21}$$

LUCY CAVENDISH COLLEGE

$$\rho \frac{d}{ds}(Av_m^2) = \rho e V_\infty \sin\chi \,, \tag{22}$$

$$\rho A \frac{v_m^2}{R} = \rho e V_\infty \cos\chi + c_d \frac{\rho}{2} V_\infty^2 b \cos^2\chi \,. \tag{23}$$

Here, A is the area of the cross-section of the jet, R is the radius of the curvature of its centre line, v_m is the jet velocity, which is assumed to be constant across every cross-section, e is the entrained volume of fluid per length of the jet and per time, c_d is the drag coefficient of the jet in the cross-flow, and b is the lateral spread of the jet.

The variation of the mass flux from one cross-section to the next one is caused by the fact, that due to the turbulent frictional forces, fluid is entrained through the surface of the jet. The variation of the momentum flux of the jet is caused by the fact, that with the entrained mass ρe also its momentum is transferred to the jet; further, the frictional forces of the crosswind act on the jet analogously as on a solid body and therefore modify the momentum flux. The influence of pressure gradients is neglected.

For the special case of vanishing crosswind, A is the area of a circle and e is a constant known from earlier investigations. A generalization for non vanishing crosswind can be gained by introduction of empirical correction functions f and g:

$$\frac{e}{(v_m - V_\infty \sin\chi)b} = e_0(1 + f(\eta)) \,, \quad \eta = \frac{V_\infty \cos\chi}{v_m - V_\infty \sin\chi} \,, \tag{24}$$

$$A = \frac{\pi}{4} b^2 \left(1 + g\left(\frac{z}{D}, \sigma\right)\right) . \tag{25}$$

For the solution of the problem, Taylor series with respect to the crosswind number $\sigma = V_\infty / V_j$ have been used. The free empirical parameters have been determined in such a way, that the theoretical results fitted experimental values of various authors.

Now, in a third step of the investigation, the reaction of the bent jet on the inviscid main flow can be treated. For this, the boundary of the jet determined in the second step is considered as the boundary of a solid body, across which a normal velocity is given as a consequence of the entrainment of the jet. For the calculation, sources on the surfaces of the plate and of the jet have been introduced, and equation (12) has been used (see [37]).

Figure 15 shows, that in the neighbourhood of the origin of the jet, the calculated curves of constant pressure coefficient c_p are situated between the measured curves of different authors, except in the wake region of the jet; that is, the agreement is very good with exception of that region.

The total turbulent viscous flow field of the jet and of the crossflow has been investigated by Oh and Schetz [26] using a finite element method. For high crosswind numbers σ and for a square orifice of the jet, this total flow problem has been solved by Rodi and Srivatsa [31] with a finite difference procedure. But it must be emphazised, that such methods consume essentially more computer time than the approximation discussed above, in which boundary elements are sufficient.

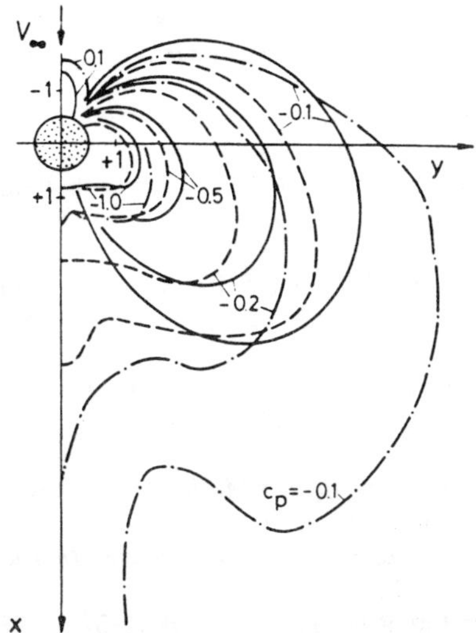

Fig. 15. Curves of constant pressure coefficient c_p on the surface of the plate at $\sigma = 0.25$; experiments:
– – – Bradbury and Wood [4],
– · – Harms and Baumert [12];
theory: —— [37]

REFERENCES

1. Barberis, D. 'Calcul de la couche limite tridimensionelle en modes direct ou inverse sur des obstacles quelconques' *Recherche Aérospatiale*, pp. 169 - 195, 1986.

2. Batchelor, G.K. *An Introduction to Fluid Dynamics* Cambridge University Press, 1967.

3. Böddener, W. 'Measurements of the Pressure Distribution on a Prolate Spheroid in the Neighborhood of a Plane Wall' *Private Communication*, 1987.

4. Bradbury, L.J.S. and Wood, M.N. 'The Static Pressure Distribution Around a Circular Jet Exhausting Normally from a Plane Wall into an Airstream', ARC C.P. 822, 1965.

5. Brebbia, C.A., Telles, J.C.F. and Wrobel, L.C. *Boundary Element Techniques* Springer-Verlag, Berlin and New York, 1984.

6. Cebeci, T., Sedlock D., Chang, K.C. and Clark, R.W. 'Analysis of Wings with Flow Separation' *J. Aircraft*, Vol.26, pp. 214 - 220, 1989.

7. Cebeci, T. and Su, W. 'Separation of Three-Dimensional Laminar Boundary Layers on a Prolate Spheroid' *J. Fluid Mech.*, Vol.191, pp. 47 - 77, 1988.

8. Coponet, D. 'Etude par vélocimétrie laser de la couche limite sur un ellipsoïde de révolution allongé du DFVLR', ONERA - PV 4/7252 AN (310/S2Ch).

9. Geißler, W. 'Three-Dimensional Laminar Boundary Layer over a Body of Revolution at Incidence and with Separation' *AIAA J.*, Vol.12, pp. 1743 - 1745, 1974.

10. Hacket, J.E. and Miller, H.R. 'The Aerodynamics of the Lifting Jet in a Cross Flowing Stream', in Analysis of a Jet in a Subsonic Crosswind, NASA SP - 218, pp. 37 - 48, *Symposium Langley Research Center*, 1969.

11. Han, T. and Patel, V.C. 'Flow Separation on a Spheroid at Incidence' *J. Fluid Mech.*, Vol.92, pp. 643 - 657, 1979.

12. Harms. L. and Baumert, W. 'Strahlinterferenz an einer Kreisscheibe', AVA-Bericht 68 A01, 1968.

13. Hinze, J.O. *Turbulence*, 2d Edition, McGraw-Hill, New York, 1975.

14. Jacob, K. 'Berechnung der Potentialströmung um Profile mit Absaugung und Ausblasen' *Ingenieur-Archiv*, Vol.32, pp. 51 - 65, 1963.

15. Jacob, K. 'Berechnung der abgelösten inkompressiblen Strömung um Tragflügelprofile und Bestimmung des maximalen Auftriebs' *Zeitschrift Flugwissen.*, Vol.17, pp. 221 - 230, 1969.
16. Jacob, K. 'Computation of the Flow Around Wings with Rear Separation', DFVLR-FB 82-22, 1982 and *J. Aircraft*, Vol.21, pp. 97 - 98, 1984.
17. Jacob, K. 'Advanced Method for Computing the Flow Around Wings with Rear Separation and Ground Effect', DFVLR-FB 86-17, 1986 and *J. Aircraft*, Vol.24, pp. 126 - 128, 1987.
18. Jacob, K. and Steinbach, D. 'A Method for Prediction of Lift for Multi-Element Airfoil Systems with Separation' in V/STOL Aerodynamics, pp. 12-1 to 12-16, *AGARD Conference Proceedings No. 143*, Delft, 1974.
19. Jaswon, M.A. 'A Review of the Theory', Chapter 1, *Topics in Boundary Element Research* ed. Brebbia, C.A. Vol.1, pp. 13 - 40, Springer-Verlag, Berlin and New York, 1984.
20. Kevorkian, J. and Cole, J.D. *Perturbation Methods in Applied Mathematics* Springer-Verlag, New York and Heidelberg, 1981.
21. Kraemer, K. 'Flügelprofile im kritischen Reynoldszahl-Bereich' *Forschung Ingenieurwesen* , Vol.27, pp. 33 - 46, 1961.
22. Lamb, H. *Hydrodynamics* 6th Edition, Dover Publications, New York, 1932.
23. Martensen, E. *Potentialtheorie* Teubner, Stuttgart, 1968.
24. Martensen, E. and v. Sengbusch, K. 'Über die Randkomponenten ebener harmonischer Vektorfelder' *Arch. Rat. Mech. Anal.*, Vol.5, pp. 46 - 75, 1960.
25. Meier, H.U. and Kreplin, H.-P. 'Experimental Study of Boundary Layer Velocity Profiles on a Prolate Spheroid at Low Incidence in the Cross Section $x_0/L = 0.64$', in Viscous and Interacting Flow Field Effects (Ed. Fiore, A.W.), pp. 169 - 189, *Proc. 5th U.S. Air Force and F.R.G. Data Exchange Agreement Meeting*, Techn. Rep. AFFDL-TR-80-3088, Wright-Patterson Air Force Base, Ohio, 1980.
26. Oh, T.S. and Schetz, J.A. 'Finite Element Simulation of Complex Jets in a Cross-flow for V/STOL Applications' *J. Aircraft*, Vol.27, pp. 389 - 399, 1990.
27. Panaras, A.G. and Steger, J.L. 'A Thin-Layer Solution of the Flow about a Prolate Spheroid' *Z. Flugw. Weltraum.*, Vol.12, pp. 173 - 180, 1988.
28. Patel, V.C. and Baek, J.H. 'Boundary Layers and Separation on a Spheroid at Incidence' *AIAA J.*, Vol.23, pp. 55 - 63, 1985.
29. Patel, V.C. and Choi, D.H. 'Calculation of Three-Dimensional Laminar and Turbulent Boundary Layers on Bodies of Revolution at Incidence', in Turbulent Shear Flows (Ed. Bradbury, L.J.S., Durst, F., Launder, B.E., Schmidt, F.W. and Whitelaw, J.H.), Vol.2, pp. 199-217, *2d Int. Symp. Turb. Shear Flows*, Imp. Col., London, 1979. Springer, New York 1980
30. Riegels, F.W. *Aerofoil Sections* Butterworth, London, 1961.
31. Rodi, W. and Srivatsa, S.K. 'A Locally Elliptic Calculation Procedure for Three-Dimensional Flows and its Application to a Jet in a Cross-Flow' *Comp. Meth. Appl. Mech. Eng.*, Vol.23, pp. 67 - 83, 1980.
32. Rotta, J.C. 'Turbulent Boundary Layer Calculations with the Integral Dissipation Method' in Computation of Turbulent Boundary Layers (Ed. Kline, S.J., Morkovin, M.V., Sovran, G. and Cockrell, D.J.), Vol.1, pp. 177 - 181, *1968 AFOSR-IFP-Stanford Conference*, Stanford, 1969.
33. Rotta, J.C. 'Fortran IV - Rechenprogramm für Grenzschichten bei kompressiblen ebenen und achsensymmetrischen Strömungen', Deutsche Luft- und Raumfahrt, FB 71-51, 1971.
34. Rotta, J.C. *Turbulente Strömungen* Teubner, Stuttgart, 1972.
35. Schlichting, H. *Boundary Layer Theory* 7th Edition, McGraw-Hill Book Company, New York, 1979.

36. Schmitt, H. 'Deflection of a Round Turbulent Jet in a Crosswind' *Arch. Mech. Stos.*, Vol.26, pp. 849 - 859, 1974.
37. Schmitt, H. 'Änderung einer Parallelströmung entlang einer ebenen Platte durch einen quer gerichteten Freistrahl' *Z. Flugwiss. Weltraum.*, Vol.3, pp. 283 - 293, 1979.
38. Schmitt, H. 'Druckverteilung an einem schräg angeströmten Rotationsellipsoid' *Ingenieur-Archiv*, Vol.53, pp. 337 - 344, 1983.
39. Schmitt, H. 'Potential and Thin-Layer Flow about a Prolate Spheroid' *Z. Flugwiss. Weltraum.*, Vol.14, pp. 117 - 119, 1990.
40. Schmitt, H. 'High Reynolds Number Flow about a Prolate Spheroid Moving near Ground', in Boundary Element Technology VI (Ed. Brebbia, C.A.), pp. 13 - 27, Computational Mechanics Publications, Southampton and Boston, 1991.
41. Schmitt, H. and Schneider, G.R. 'Einfluß des Auftriebs eines angestellten Rotationsellipsoids auf die Grenzschichtströmung in der Symmetrieebene' *ZAMM*, Vol.68, pp. T353 - T355, 1988.
42. Schmitt, H. and Schneider, G.R. 'Calculation of the Potential Flow with Consideration of the Boundary Layer' Chapter 6, *Topics in Boundary Element Research* ed. Brebbia, C.A. Vol.5, pp. 110 - 133, Springer-Verlag, Berlin and New York, 1989.
43. Steinbach, D. 'Berechnung der Strömung mit Ablösung für Profile und Profilsysteme in Bodennähe oder in geschlossenen Kanälen' *Z. Flugwiss. Weltraum.*, Vol.2, pp. 293 - 305, 1978.
44. Tanaka, I. 'Three-Dimensional Ship Boundary Layer and Wake' *Advances in Applied Mechanics* ed. Hutchinson, J.W. and Wu, T.Y., Vol.26, pp. 311 - 359, Academic Press, Boston and New York, 1988.
45. Van Dyke, M. *An Album of Fluid Motion* Parabolic Press, Stanford, 1982.
46. Wang, K.C. 'Boundary Layer over a Blunt Body at Low Incidence with Circumferential Reversed Flow' *J. Fluid Mech.*, Vol.72, pp. 49 - 65, 1975.

The Complete Double Layer Boundary Integral Equation Method for Particles Moving Close to Boundaries

H. Power, B. Febres de Power

Instituto de Mecánica de los Fluidos, Universidad Central de Venezuela, Caracas, Venezuela

ABSTRACT

The problem of determining the slow motion of a particle of arbitrary shape near a plane boundary, rigid or not, in a viscous fluid is formulated exactly as a system of linear Fredholm integral equations of the second kind, by completing the deficient range of a double layer potential and using the adequate image system needed to satisfy the boundary or matching conditions at the boundary. It is shown that this system of integral equations possesses a unique continuous solution when the boundary of the particle is a Lyapunov surface and the velocity data on the boundary surface is continuous and this system is used as the basis of a numerical model that uses standard boundary element techniques.

INTRODUCTION

To deal with the motion of bodies of arbitrary shape near a boundary it is very attractive to use integral equation approach. Integral representations formulae analogous to those employed in potential theory exist for Stokes flow and their use can be traced back to the work of Lorentz [1], an intensive use of these integral representations formulae in the numerical solution of Stokes' flows have been applied since the work done by Youngren and Acrivos [2], they showed how integral equations of the first kind for the surface tractions on a particle in arbitrary unbounded flow could be numerically solved in practice. Power et al. [3] give an exact formulation for the slow viscous flow due to the arbitrary motion of a particle of arbitrary shape near a plane rigid wall, these formulation is found in terms of a system of linear Fredholm integral equations of the first kind for the distribution over the particle surface of a Stokeslet (single layer) and its image system needed to satisfy the no-slip boundary condition at the wall, this integral equation system has a unique solution. They also developed a numerical method based on these integral equations and tested them it for the cases with known analytical solution. Recently, Hsu and Ganatos [4] developed a first kind integral equation method similar to the previous one, but they used the complete Green's integral representation formulae instead of only a single layer potential. Contrasting with Youngren and Acrivos' work for an unbounded domain, where the double layer potential term vanishes when the velocity boundary datum corresponds to a rigid body motion velocity, in Hsu and Ganatos representation formulae, the image part of the double layer does not vanish for a rigid body motion boundary velocity, and therefore it is necessary to carry out an extra numerical work, when it is compared with the single layer representation alone. An apparent advantage of the Green's formulation is that the unknown density can be identified with the local stress forces and in the single layer representation the density does not have any direct physical meaning. However, Power et.al. [3] proved that in spite of the fact that the single layer density is not equal to the local surface stress forces, their corresponding surface integration are identical and thus equal to the total force . Later, Power et.al [5] using an analysis similar to their previous one determined the slow viscous flow due to the arbitrary motion of a particle of arbitrary shape near a plane interface, in this case they used the image system needed to satisfy the matching conditions at a plane fluid interface between two fluids with diferent viscosities and densities (for a good survey of the image method used in conjunction with the boundary integral equation method see Weinbaum and Ganatos [6]).

As it is known, Fredholm integral equations of the first kind generally give rise to unstable numerical schemes based upon discretization for the surface integrals involved (see Goldberg [7] page 36), the instability manifesting itself in the ill-conditioning of the matrix approximation of the kernel. Nonetheless, it is possible to apply the discretization method if only low-order accuracy is desired and the system of linear equations that needs to be solved is not too ill-conditioned, as appears to be the case in those works that use Youngren and Acrivos method. On the other hand, solving an equation of the second kind is a well-posed problem (fort a good argument why integral equations of the second kind should be better than integral equations of the first kind see Karrila and Kim [8]).

Power and Miranda [9] explained how integral equations of the second kind can be obtained for general unbounded Stokes flows about a single particle. They observed that although the double layer representation, that originates a second kind integral equation coming from the jump property of the velocity field of a double layer potential across the density carrying surface, can represent only those flow fields that correspond to a force and torque free surface, the representation may be completed by adding terms that give arbitrary total force and torque in suitable linear combination, precisely a Stokeslet and Rotlet located in the interior of the particle. The extension of Power and Miranda's method to multiple particles in an unbounded flow was given by Power [10] and to a particle in a flow bounded by an exterior container by Power and Miranda [11].

Karrila and Kim [8] and Karrila, Fuentes and Kim [12] give an elegant mathematical interpretation of the above method for multiple particles problems in and unbounded flow and bounded by an exterior container. They observe that this method relates to Wielandt's deflation, by removing the end points of the spectrum of the integral operator of an integral equation of the second kind coming from a double layer representation without any completion, those eigenvalues are moving to the origin without affecting the rest of the eigenvalues. Such modification will allow direct iterative solution. Karrila and Kim called the new method the *Completed Double Layer Boundary Integral Equation Method*, since it involves the idea of completing the deficient range of the double layer operator. A very important contribution of these two papers is that they show that the method is the most efficient one to solve numerically the mobility problem, where the force and torque on each particle is specified and the unknown particle motion is to be determined.

The main objective of this paper is to extend Power and Miranda's completed double layer boundary integral equation method to the problem of particles of arbitrary shape moving arbitrarily near a plane boundary, rigid or not, by completing the deficient range of the double layer potential and using the adequate image system.

2) PROPOSED FORM OF SOLUTION

We consider the problem of determining the low Reynolds number incompressible viscous flow due to the motion of a solid particle of arbitrary shape with boundary surface S of Lyapunov type (see Gunter [13]) near a plane boundary π which we shall take to be $x_3 = 0$, the geometric center of the particle will be chosen to be at $x_1 = 0$, $x_2 = 0$ and $x_3 = h$. Under this condition, the velocity field and pressure $(\vec{u}, p)$, satisfy as a first approximation the Stokes' system of equations, where for simplicity, we take the dynamic viscosity $\mu = 1$, since the general case can always be reduced to this one by a coordinate transformation:

$$\begin{aligned} \frac{\partial^2 u_i(x)}{\partial x_j \partial x_j} &= \frac{\partial p(x)}{\partial x_i} \qquad x \in \Omega_e \\ \frac{\partial u_i(x)}{\partial x_i} &= 0 \qquad\qquad x \in \Omega_e \end{aligned} \tag{2.1}$$

here Ω_e is the domain above π exterior to the particle having S and π as boundaries, Ω_i will designate the domain interior to the particle.

The velocity field satisfies the non–slip boundary condition on S:

$$u_i(\xi) = U_i(\xi) \qquad \text{for all } \xi \in S, \tag{2.2}$$

and the corresponding boundary or matching conditions on the plane π, as well as the following conditions at infinity

$$u_i = 0(R^{-1}), \qquad p = 0(R^{-2}) \qquad \text{as } |\, x \,| \to \infty \tag{2.3}$$

where R is the distance from the geometric center of the particle to a point x in the flow field, and $\vec{U}(x)$ is the prescribed velocity at the surface of the particle.

Following Power and Miranda's [9] completed method, we will seek the solution for the velocity field of the above boundary value problem in the following form:

$$\begin{aligned} u_i(x) = \int_S K_{ij}(x,y)\phi_j(y)\, dS_y + \int_S K^*_{ij}(x,y^*)\phi_j(y)\, dS_y + S_i^j(x,y^0)\alpha_j + \\ S_i^{*j}(x,y^{0*})\alpha_j + R_i^j(x,y^0)w_j + R_i^{*j}(x,y^{0*})w_j \end{aligned} \tag{2.4}$$

where y^0 is the point $(0,0,h)$ and y^{0*} is the point $(0,0,-h)$, here

$$S_i^j(x,y) = \frac{1}{8\pi}\left(\frac{\delta_{ij}}{r} + \frac{(x_i - y_i)(x_j - y_j)}{r^3}\right) \qquad (2.5\text{-a})$$
$$r = \mid x - y \mid$$

is the fundamental singular solution of Stokes' equations, known as "Stokes-let" located at the point y and oriented in the j-th direction,

$$\begin{aligned} K_{ij}(x,y) &= \sigma_{ij}(\vec{S}^k(x,y))_y n_k(y) \\ &= -\frac{3}{4\pi}\frac{(x_i - y_i)(x_j - y_j)(x_k - y_k)n_k(y)}{r^5} \end{aligned} \qquad (2.5\text{-b})$$

is the symmetric component of a Stokes doublet, which is, in itself, a fundamental singularity called Stresslet, where

$$\sigma_{ij}(\vec{u}^k(x,y))_x = -p^k\delta_{ij} + \left(\frac{\partial u_i^k}{\partial x_j} + \frac{\partial u_j^k}{\partial x_i}\right), \qquad (2.5\text{-c})$$

is the stress tensor corresponding to the flow field $(\vec{u}^k, p^k)$,

$$R_i^j(x,y) = \frac{1}{8\pi}\frac{\varepsilon_{ilk}\delta_{lj}x_k}{r^3} \qquad (2.5\text{-d})$$

is a singular solution of the Stokes' nonhomogeneous equation with a forcing function equal to $\varepsilon_{ilk}(\partial/\partial x_k)\delta_{lj}\delta(x)$, singularity called Rotlet.

In equation (2.4), $S_i^{*j}(x, y^{0*})$, and $R_i^{*j}(x, y^{0*})$ are the corresponding image system of the Stokeslet and Rotlet placed at the point y^{0*}, needed to satisfy the boundary or matching conditions at the plane π, and $K_{ij}^*(x, y^*)$ is the image system of the Stresslet $K_{ij}(x,y)$.

It can be observed that, in equation (2.4), we have added to the double layer potential, which is the distribution of Stresslet, $K_{ij}(x,y)$ over the particle surface, a image double layer, coming from the distribution of the Stresslet image system, $K_{ij}^*(x,y^*)$, with $y_1^* = y_1$, $y_2^* = y_2$ and $y_3^* = -y_3$, over the particle surface with the same unknown density $\vec{\phi}(y)$ of the double layer, a Stokeslet, $\vec{S}^j(x, y^0)$, located at the center of the particle, and its corresponding image system, $\vec{S}^{*j}(x, y^{*0})$, both of them with strength equal to the constant vector $\vec{\alpha}$, and a Rotlet, $\vec{R}^j(x, y^0)$, located at the center of the particle and its corresponding image system, $\vec{R}^{*j}(x, y^{*0})$, also both of them with the same vector strength $\vec{w}$ and therefore this representation formulae automatically satisfies the boundary or matching conditions at the plane π.

It is convenient for later use to choose $\vec{\alpha}$ and $\vec{w}$ as follows:

$$\alpha_i = \int_S \phi_j(\xi)\varphi_j^i(\xi)\, dS_\xi \qquad \text{for } i = 1,2,3 \qquad (2.6\text{-a})$$

$$w_i = \int_S \phi_j(\xi)\varphi_j^{i+3}(\xi)\, dS_\xi \qquad \text{for } i = 1,2,3 \qquad (2.6\text{-b})$$

where $\vec{\varphi}^i$ for $i = l, 2, \cdots 6$ give the motion of the fluid as a rigid body, taken to be:

$$\vec{\varphi}^i(x) = (\delta_{1i}, \delta_{2i}, \delta_{3i}) \qquad \text{for } i = 1,2,3$$
$$\vec{\varphi}^4(x) = (0, x_3, -x_2), \qquad \vec{\varphi}^5(x) = (-x_3, 0, x_1), \qquad \vec{\varphi}^6(x) = (x_2, -x_1, 0) \tag{2.7}$$

As is well known, a Stokeslet exerts a total force equal to its strength and zero total torque on any closed surface enclosing it, and a Rotlet exerts a total torque equal to its strength and zero total force on any surface enclosing it. Therefore, since a hydrodynamic double layer potential with a vector tension having a well defined limiting value at points of S yields zero total force and torque on S, and since all the image systems used in equation (2.4) are regular Stokes flow in the domain above the plane π, because they are the superposition of singular solutions located outside these domain, and consequently does not exert total force and torque on S, it can be concluded that the total force and torque resulting from the flow field defined by equation (2.4) will be equal to $\vec{\alpha}$ and $\vec{w}$ respectively.

From now on we will assume that $\vec{U}$ is continuous on S. Applying the boundary condition (2.2) to the flow field defined by the expressions (2.4) leads to the following linear system of Fredholm integral equations of the second kind for the unknown vector density $\vec{\phi}$:

$$\begin{aligned} U_i(\xi) = -\frac{1}{2}\phi_i(\xi) + \int_S K_{ij}(\xi, y)\phi_j(y)\, dS_y + \int_S K^*_{ij}(\xi, y^*)\phi_j(y)\, dS_y + \\ S^j_i(\xi, y^0)\alpha_j + S^{*j}_i(\xi, y^{0*})\alpha_j + \\ R^j_i(\xi, y^0)w_j + R^{*j}_i(\xi, y^{0*})w_j \qquad \xi \in S \end{aligned} \tag{2.8}$$

To show that (2.8) possesses a unique continuous solution $\vec{\phi}$ for continuous datum $\vec{U}$ it is sufficient according to Fredholm's alternative to show that the following homogeneous system (2.9) for $\vec{\phi}^0$ admits only the trivial solution in the space of continuous functions:

$$\begin{aligned} 0 = -\frac{1}{2}\phi^0_i(\xi) + \int_S K_{ij}(\xi, y)\phi^0_j(y)\, dS_y + \int_S K^*_{ij}(\xi, y^*)\phi^0_j(y)\, dS_y + \\ S^j_i(\xi, y^0)\alpha^0_j + S^{*j}_i(\xi, y^{0*})\alpha^0_j + \\ R^j_i(\xi, y^0)w^0_j + R^{*j}_i(\xi, y^{0*})w^0_j \qquad \xi \in S \end{aligned} \tag{2.9}$$

where

$$\alpha^0_i = \int_S \phi^0_j(\xi)\varphi^i_j(\xi)\, dS_\xi \qquad \text{for } i = 1,2,3 \tag{2.10-a}$$

$$w^0_i = \int_S \phi^0_j(\xi)\varphi^{i+3}_j(\xi)\, dS_\xi \qquad \text{for } i = 1,2,3 \tag{2.10-b}$$

To prove that (2.9) admits only the trivial solution we will follow the ideas given by Power and Miranda [9] for the solution of the analog unbounded problem to the one worked here. From equations (2.9) it follows that the pair of vector fields $\vec{V}_1(x)$, $\vec{V}_2(x)$ defined by (2.11) and (2.12) below, which are Stokes' velocity fields in Ω_e bounded at infinity, coincide on S:

$$V_i^1(x) = \int_S K_{ij}(x,y)\phi_j^0(y)\,dS_y + \int_S K_{ij}^*(x,y^*)\phi_j^0(y)\,dS_y + R_i^j(x,y^0)w_j^0 + R_i^{*j}(x,y^{0*})w_j^0 \tag{2.11}$$

$$V_i^2(x) = -(S_i^j(x,y^0)\alpha_j^0 + S_i^{*j}(x,y^{0*})\alpha_j^0) \tag{2.12}$$

Since $\vec{V}^1$ and $\vec{V}^2$ are regular Stokes' flows in Ω_e having the same boundary value on S and satisfy the same boundary or matching conditions at π, then by the uniqueness of solutions of the present problem, it follows that $\vec{V}^1$ and $\vec{V}^2$ are identically equal in Ω_e; but since $\vec{V}^1(x) = O(R^{-2})$ as $\mid x \mid \rightarrow \infty$ and $\vec{V}^2(x) = O(R^{-1})$, it follows that both $\vec{V}^1$ and $\vec{V}^2$ are identically zero in Ω_e, i.e.

$$\int_S K_{ij}(x,y)\phi_j^0(y)\,dS_y + \int_S K_{ij}^*(x,y^*)\phi_j^0(y)\,dS_y + R_i^j(x,y^0)w_j^0 + R_i^{*j}(x,y^{0*})w_j^0 = 0 \qquad \text{when } x \in \Omega_e \tag{2.13}$$

and

$$\alpha_i^0 = \int_S \phi_j^0(\xi)\varphi_j^i(\xi)\,dS_\xi = 0 \qquad \text{for } i = 1,2,3 \tag{2.14}$$

On the other hand (2.13) implies that the pair $\vec{V}^3$ and $\vec{V}^4$ of Stokes' velocity fields defined by (2.15) and (2.16) below are identically equal in Ω_e:

$$V_i^3(x) = \int_S K_{ij}(x,y)\phi_j^0(y)\,dS_y + \int_S K_{ij}^*(x,y^*)\phi_j^0(y)\,dS_y \tag{2.15}$$

$$V_i^4(x) = -(R_i^j(x,y^0)w_j^0 + R_i^{*j}(x,y^{0*})w_j^0) \tag{2.16}$$

Since the torque resulting from $\vec{V}^4$ on S is equal to $-\vec{w}^0$ and the torque due to a double layer potential and its image system is equal to zero when it is well defined, it can be concluded that:

$$w_i^0 = \int_S \phi_j^0(\xi)\varphi_j^{i+3}(\xi)\,dS_\xi = 0 \qquad \text{for } i = 1,2,3 \tag{2.17}$$

and

$$V_i^3(x) = \int_S K_{ij}(x,y)\phi_j^0(y)\,dS_y + \int_S K_{ij}^*(x,y^*)\phi_j^0(y)\,dS_y = 0 \tag{2.18}$$

for every $x \in \Omega_e$. Therefore equation (2.9) reduces to:

$$-\frac{1}{2}\phi_i^0(\xi) + \int_S K_{ij}(\xi,y)\phi_j^0(y)\,dS_y + \int_S K_{ij}^*(\xi,y^*)\phi_j^0(y)\,dS_y = 0 \tag{2.19}$$

for every $\xi \in S$

The above homogeneous equation have precisely six linearly independent solutions $\vec{\phi}^{0k}$, for $k = 1,2,\cdots,6$, defined by the previous given rigid body motion vectors. To prove it, we will use an approach suggested by Karrila and Kim [14] when they show that the six rigid body motion are the eigenfunctions corresponding to the homogeneous equation coming from an exterior double layer potential alone, without the image system. Karrila and Kim's proof does not use a single layer potential as it is used in the classical proof given by Ladyzhenskaya [15]. Equation (2.19) shows that the double layer potential with density $\vec{\phi}^0$ plus its image system needed to satisfy the boundary or matching conditions at the plane π, which is a regular Stokes flow field in Ω_e, has zero velocity value at every point ξ belonging to the surface S. Therefore, from the uniqueness of solution of the present problem, it follows that this combination of double layer and its image system, has to be zero for every point x belonging to Ω_e and the same has to be true for the vector tension, $\sigma_{ij}n_j$, at every point in the domain Ω_e and in particular at points $\xi \in S$. Since the double layer potential has continuous vector tension across its density carrying surface and the image system of the double layer potential used here is regular at every point above the plane π, it follows that the zero vector tension of this combination can be extended continuously across the density carrying surface S, and therefore inside the particle, this flow combination has to behave as a rigid body motion. On the other hand, from the jump property of the velocity of a double layer potential across its density carrying surface, we obtain:

$$(W_i(\xi) + W_i^*(\xi))_{(i)} - (W_i(\xi) + W_i^*(\xi))_{(e)} = \phi_i^0(\xi)$$

the subscript (i) denotes the limiting value coming from the inside of the particle and the subscript (e) the limiting value coming from the outside of the particle. Therefore, the eigenfunction $\vec{\phi}^0$ has to be equal to the six rigid body motion vectors $\varphi_i^k(\xi)$ with $k = 1, 2, \cdots 6$, since from the above statement, we can conclude that $(W_i(\xi)+W_i^*(\xi))_{(e)} = 0$ and $(W_i(\xi)+W_i^*(xi))_{(i)} = \varphi_i^k(\xi)$ with $k = 1, 2, \cdots, 6$, then necessarily $\phi_i^0 = \sum_{k=1}^{6} C_k\varphi_i^k$ for $i = 1, 2, 3$, where $C_1, C_2, \cdots, C_6$ are some real constants.

Equations (2.14) and (2.17) imply for $i = 1, 2, 3$ that:

$$C_k \int_S \varphi_j^k(y)\varphi_j^i(y)\, dS_y = 0$$

and

$$C_k \int_S \varphi_j^k(y)\varphi_j^{i+3}(y)\, dS_y = 0 \tag{2.20}$$

The above linear algebraic system for $C_1, C_2, \cdots C_6$ only admits the trivial solution, implying that $\vec{\phi}^0 = (0,0,0)$ on S, because the determinant of (2.20) has

$$\int_S \varphi_j^l(y)\varphi_j^q(y)\, dS_y$$

as element in the lth row and q^{th} column, and is thus the Gram determinant for the vector functions $\vec{\varphi}^1, \vec{\varphi}^2, \cdots \vec{\varphi}^6$ with a non vanishing value, on account of the linear independence of $\vec{\varphi}^k, k = 1, 2, \cdots 6$.

We have established that the non homogeneous integral equation (2.8) has a unique continuous solution $\vec{\phi}$ for any arbitrary continuous datum $\vec{U}$, and the

Stokes' velocity field defined with this $\vec{\phi}$, by equations (2.4) and (2.6), with its corresponding pressure, provides the solution of the boundary value problem defined by equations (2.1) through (2.3).

From the stated dynamic properties of the singularities here considered as well as those of the double layer potential, it is found that the total force and torque exerted on S by the the flow field represented by (2.4) trough (2.6) are:

$$\begin{aligned} F_i = \alpha_i &= \int_S \phi_j(\xi)\varphi_j^i(\xi)\, dS_\xi \qquad \text{for } i = 1,2,3 \\ T_i = w_i &= \int_S \phi_j(\xi)\varphi_j^{i+3}(\xi)\, dS_\xi \qquad \text{for } i = 1,2,3 \end{aligned} \tag{2.21}$$

3) NUMERICAL SOLUTION

An apparent difficulty in the numerical solution of (2.8) is that the integrands appearing in the integrals $\int_S K_{ij}(\xi, y)\, \phi_j(y)\, dS_y$ are singular at $y = \xi$. One way around this difficulty is to write equation (2.8) in the following way:

$$\begin{aligned} U_i(\xi) = &\int_S K_{ij}(\xi, y)(\phi_j(y) - \phi_j(\xi))\, dS_y + \int_S K_{ij}^*(\xi, y^*)\phi_j(y)\, dS_y + \\ &S_i^j(\xi, y^0)\alpha_j + S_i^{*j}(\xi, y^{0*})\alpha_j + \\ &R_i^j(\xi, y^0)w_j + R_i^{*j}(\xi, y^{0*})w_j \qquad \xi \in S \end{aligned} \tag{3.1}$$

for $i = 1,2,3$, which can be done since

$$\int_S K_{ij}(\xi, y)\, dS_y = \frac{1}{2}\,\delta_{ij}$$

Now

$$K_{ij}(\xi, y) = -\frac{3}{4\pi}\frac{(\xi_i - y_i)(\xi_j - y_j)(\xi_k - y_k)}{r^5}\, n_k(y) = -\frac{3}{4\pi}\frac{\partial r}{\partial \xi_i}\frac{\partial r}{\partial \xi_j}\frac{\partial}{\partial n(y)}\left(\frac{1}{r}\right)$$

Then

$$\int_S (\phi_j(y) - \phi_j(\xi))K_{ij}(\xi, y)\, dS_y = -\frac{3}{4\pi}\int_S (\phi_j(y) - \phi_j(\xi))\frac{\partial r}{\partial \xi_i}\frac{\partial r}{\partial \xi_j}\, d\Omega \tag{3.2}$$

Since $(\partial/\partial n(y))(1/r)dS_y$ happens to be $d\Omega$, that is, the differential of the solid angle at point y, then the integrals above are proper under the new variable of integration. It follows that the singularities in (3.1) can be simply removed by setting their integrands equal to zero when $y = \xi$.

For a general Lyapunov surface S, the numerical solution of (3.1) can be done by a direct discretization of the surface integral involved, transforming it into a linear system of algebraic equations. This is accomplished by dividing S into N–elements Δ_j, $j = 1,2,\cdots,N$, are small in respect to S and over which the unknown vector density of $\vec{\phi}$ may be considered approximately constant and equal to their value at the center of the element. Then, we can write (3.1) approximately as:

$$
\begin{aligned}
U_i(\xi^m) \sim \sum_{l=1}^{N} \{ & (\phi_j(\xi^l) - \phi_j(\xi^m))\,(1-\delta_{lm})A^l_{ij} + \phi_j(\xi^l)\;A^{l*}_{ij} + \\
& (S^j_i(\xi^m, y^0) + S^{*j}_i(\xi^m, y^{0*}))\phi_k(\xi^l)B^l_{kj} + \\
& (R^j_i(\xi^m, y^0) + R^{*j}_i(\xi^m, y^{0*}))\phi_k(\xi^l)D^l_{kj} \} \\
& i = 1,2,3 \text{ and } k = 1,2,\cdots,N
\end{aligned}
\tag{3.3}
$$

where the terms $(\phi_j(\xi^l) - \phi_j(\xi^m))\,(1-\delta_{lm})A^l_{ij}$ are equal to zero when $l = m$ as was pointed out, due to the removal of singularities, and

$$
\begin{aligned}
A^l_{ij} &= \int_{\Delta_l} K_{ij}(\xi^m, y)\; dS_y \qquad \text{for } l \neq m \\
A^{l*}_{ij} &= \int_{\Delta_l} K^*_{ij}(\xi^m, y^*)\; dS_y \\
B^l_{ij} &= \int_{\Delta_l} \varphi^i_j(y)\; dS_y \qquad \text{for } i = 1,2,3 \\
D^l_{ij} &= \int_{\Delta_l} \varphi^{i+3}_j(y)\; dS_y \qquad \text{for } i = 1,2,3
\end{aligned}
$$

This algebraic system can be solved numerically using suitable integration and matrix inversion techniques, here all integrals involved are proper, for the previously stated removal of singularities. Therefore, the numerical calculation of these integrals presents no problem. This removal of singularities is another advantage of the second kind formulation.

After finding the local density $\phi_i(\xi^m)$, for $m = 1,2,\cdots,N$ and $i = 1,2,3$, the total force upon the surface S is given by equation (2.21). The numerical integration used in this work has been done with Gaussian quadratures and the linear algebraic systems were solved using Gaussian elimination followed by iterative improvement (For more details about the numerical calculation of integral equation (3.1) see the paper by Power and Miranda [9]).

The hydrodynamic interaction between a body of simple shape, like spheres, spheroids, etc., with a plane wall at small Reynolds number has been the subject of many studies (for a good literature survey see Chapter 7 of Happel and Brenner[16]). The motion of a sphere parallel to a plane wall was treated by Faxen [17], using the method of reflections; this method amounts to seek a systematic scheme of successive iterations by which the boundary value problem may be solved by considering boundary conditions associated with one boundary at a time. Faxen obtained that the force on the particle lies along the direction of motion and given by the following expression:

$$
|\vec{F}| = \frac{6\pi\mu a U}{(1 - (9/16)(a/L) + (1/8)(a/L)^3 - (45/256)(a/L)^4 - (1/6)(a/L)^5)}
\tag{3.4}
$$

(which includes the effects of a second reflection). Here μ is the fluid viscosity, a is the radius of the sphere, L is the distance between the center of the sphere and the wall, and U is the magnitude of the velocity of the sphere. O'Neill [18] solved exactly the translation motion of a sphere parallel to a wall using the general bipolar coordinate solution of the creeping motion equation employed by Stimson and Jeffery [19] in their solution of the problem of two spheres falling along their line of centers. The resistance due to this motion was computed numerically by Goldman et al. [20] using O'Neill's series solution and given here in Figure 1.

The problem of a sphere approaching perpendicularly a plane wall was solved exactly by Brenner [21] using the general bipolar coordinate solution and given here in Figure 2. A treatment by Wakiya [22], using the method of reflections gives the resistance due to the motion of a single sphere toward a plane wall to $O(a/L)^3$ as:

$$|\vec{F}| = \frac{(6\pi\mu a U)}{(1-(9/8)(a/L)+(1/2)(a/L)^3)}. \tag{3.5}$$

At a dimensionless distance $L/a > 10$, this formula agrees closely with the values computed using Brenner's relation.

In order to test the numerical method developed in this work, valid for arbitrary particle shape and arbitrary plane boundary, rigid or not, as long the image system needed to satisfy the boundary or matching conditions at the boundary is known, we solve the flow due to translation motion of a sphere parallel and perpendicular to a rigid plane wall for different values of the local dimensionless distance $L^* = L/a$ (the image systems of the Stokeslet, Rotlet and Stresslet needed to satisfy the zero velocity condition on the plane wall π are given in the appendix). Comparisons of the above numerical solution with the analytical solutions previously presented are given in figures 1 and 2. Our numerical calculations have been performed in a Personal Computer 6Mb. RAM, 25MHz Intel 80386 CPU with mathematical coprocessor, under this condition the maximum number of elements tested was $N = 96$.

APPENDIX:

The image system of the Stokeslet needed to satisfy the non–slip boundary condition at the wall, was found by Blake [23] and it is equal to:

$$S_i^{*j}(x,y^*) = -S_i^j(x,y^*) + 2h(\delta_{j\alpha}\delta_{\alpha k} - \delta_{j3}\delta_{3k})(hD_i^k(x,y^*) - SD_{il}^k(x,y^*)\delta_{l3}) \tag{A-1}$$

here the tensor $(\delta_{j\alpha}\delta_{\alpha k} - \delta_{j3}\delta_{3k})$ is not zero only when $j = k$; its value is $+1$ for $j = 1$ or 2, and -1 for $j = 3$,

$$D_i^k(x,y^*) = \frac{1}{8\pi}\frac{\partial}{\partial x_k}\left(\frac{x_i - y_i^*}{(r^*)^3}\right)$$
$$r^* = \mid x - y^* \mid \tag{A-2}$$

is the velocity field due to a Source doublet placed at the point y^* and oriented in the k-th direction,

$$SD_{ij}^k(x,y^*) = \frac{\partial}{\partial x_k}(S_i^j) \tag{A-3}$$

is the velocity field due to Stokes doublet oriented in the k-th direction, coming from the derivative of a Stokeslet placed at the point y^* and oriented in the j-th direction.

The image system of the Rotlet needed satisfy the non–slip boundary condition at the wall, was found by Blake and Chwang [24], and is given by:

$$R_i^{*j}(x,y^*) = -R_i^j(x,y^*) + 2h\varepsilon_{kj3}D_i^k(x,y^*) + \varepsilon_{kj3}\sigma_{ik}(\vec{S}^l(x,y^*))_x\delta_{l3} \tag{A-4}$$

Pozrikidis [25] showed that $\sigma_{ij}(\vec{S}^k(x,y) + \vec{S}^{*k}(x,y^*))_y g_{jk}$, for any constant tensor g_{jk}, is a legitimate Stokes' velocity field, with vanishing value along the wall π. Therefore, since $\sigma_{ij}(\vec{S}^k(x,y))_y n_k(y) = K_{ij}(x,y)$, it should be that $K_{ij}^*(x,y^*) = \sigma_{ij}(\vec{S}^{*k}(x,y^*))_y n_k(y)$ then

$$W_i^*(x) = \int_S K_{ij}^*(x,y^*)\phi_j(y)\, dS_y \tag{A-5}$$

is the image system of the double layer potential

$$W_i(x) = \int_S K_{ij}(x,y)\phi_j(y)\, dS_y \tag{A-6}$$

REFERENCES

[1] Lorentz, H.A.: Ein Allgemeiner Satz, die Bewegung einer Reibenden Flussigkeit Betreffend, nebst einegen Anwendungen desselben (A General Theorem Concerning the Motion of a Viscous Fluid and a Few Consequences Dervived from it), *Versl. Kon. Akad. Wetensch.*, Vol. 5. (1896)

[2] Youngreen, G.K. and Acrivos, A.: Viscous Flows Past a Spheroid, *J. Fluid Mech.*, Vol. 69. (1975)

[3] Power, H.; Miranda, G. and Gonzalez, R.: Integral Equation Solution for the Flow Due to the Motion of a Body of Arbitrary Shape near a Plane Wall at Small Reynolds Number, *Math. Aplic. Comp.*, Vol. 4. (1985)

[4] Hsu, R. and Ganatos, P.: The Motion of a Rigid Body in Viscous Fluid Bounded by a Plane Wall, *J. Fluid Mech.*, Vol. 207. (1989)

[5] Power, H., Garcia, R. and MIranda, G.: Integral Equation Solution for the Flow Due to the Motion of a Body of Arbitrary Shape near a Plane Interface at Small Reynolds Number, *Applied Numerical Math.*, Vol. 2. (1986)

[6] Weinbaum, S. and Ganatos, P.: Numerical Multipole and Boundary Integral Equation Techniques in Stokes Flow, *Annu. Rev. Fluid Mech.*, Vol. 22. (1990)

[7] Goldberg, M.A.: Solution Methods for Integral Equations. Theory and Aplications, *Plenum Press*, New York. (1978)

[8] Karrila, S.J. and Kim, S.: Integral Equations of the Second Kind for Stokes Flow: Direct Solution for Physical Variables and Removal of Inherent Accuracy Limitations, *Chem. Eng. Commun.*, Vol. 82. (1989)

[9] Power, H. and Miranda, G.: Second Kind Integral Equation Formulation of Stokes Flows Past a Particle of Arbitrary Shape, *SIAM Appl.*, Vol. 47. (1987)

[10] Power, H.: Second Kind Integral Equation Solution of Stokes Flows Past n Bodies of Arbitrary Shapes, 9th Int. Conf. on BEM, Sttutgard, Computational Mechanics Publications, *Southampton and Springer Verlag*, Berlin. (1987)

[11] Power, H. and Miranda, G.: Integral Equation Formulation for the Creeping Flow of an Incompressible viscous Fluid between Two Arbitrarily Closed Surfaces and a Possible Mathematical Model for the Brain Fluid Dynamics, *J. Math. Anal. and Appl.*, Vol. 137. (1989)

[12] Karrila, S.J.; Fuentes, Y.O. and Kim, S.: Parallel Computational Strategies for Hydrodynamic Interactions between Rigid Particles of Arbitrary Shape in a Viscous Fluid, *J. Rheolog.*, Vol . 33. (1989)
[13] Gunter, N.M.: Potential Theory and its Applications to Basic Problems, *Frederick Ungar Publishing*, New York. (1967)
[14] Karrila, S.J. and Kim, S.: Foundations of Parallel Computational Microhydrodynamics: The Completed Double Layer Boundary Integral Equation Method, *Univ. of Wisconsin–Madison*, RRC123. (1991)
[15] Ladyzhenskaya, O.A.: The mathematical theory of viscous incompressible flow, *Gordon and Breach*, New York. (1963)
[16] Happel, J. and Brenner, H.: Low Reynolds Number Hydrodynamics with Spetial Applications to Particle Media, *Noordhoff International Publishing*, Netherland. (1973)
[17] Faxen, H.: Die Bewegung einer starren Kugel Langs der Achse eines mit zaner Flusigkeit gefullten Rohres, *Arkiv. Mat. Astron. Fys.*, Vol. 17. (1923)
[18] O'Neill, M.E.: Slow Motion of Viscous Liquid Caused by a Slowly Moving Body, *Mathematika*, Vol. 11. (1964)
[19] Stimson, M. and Jeffery, G.B.: The Motion of Two Spheres in a Viscous Fluid, *Proc. Roy. Soc.*, Vol. A111. (1926)
[20] Goldman, A.J.; Cox, R.G. and Brenner H.: Slow viscous motion of a sphere parallel to a plane wall, I. Motion through a quiscent fluid, *Chem. Eng. Sci.*, Vol. 22. (1967)
[21] Brenner, H.: The Slow Motion of a Sphere through a viscous fluid towards a plane surface, *Chem. Eng. Sci.*, Vol. 16. (1961)
[22] Wakiya, S.J.: Research Report 9, *Fac. Eng. Niigata Univ.*, Japan. (1960)
[23] Blake, J.R.: A Note on the Image System for a Stokeslet in a No-Slip Boundary, *Proc. Cambridge Philos. Soc.*, Vol. 70. (1971)
[24] Blake, J.R. and Chwang A.T.: Fundamental Singularities of Viscous Flow: The Image Systems in the Vecinity of a Stationary No–Slip Boundary, *J. Eng. Math.*, Vol. 8. (1974)
[25] Pozrikidis, C.: The Deformation of a Liquid Drop Moving Normal to a Plane Wall, *J. Fluid Mech.*, Vol. 215. (1990)

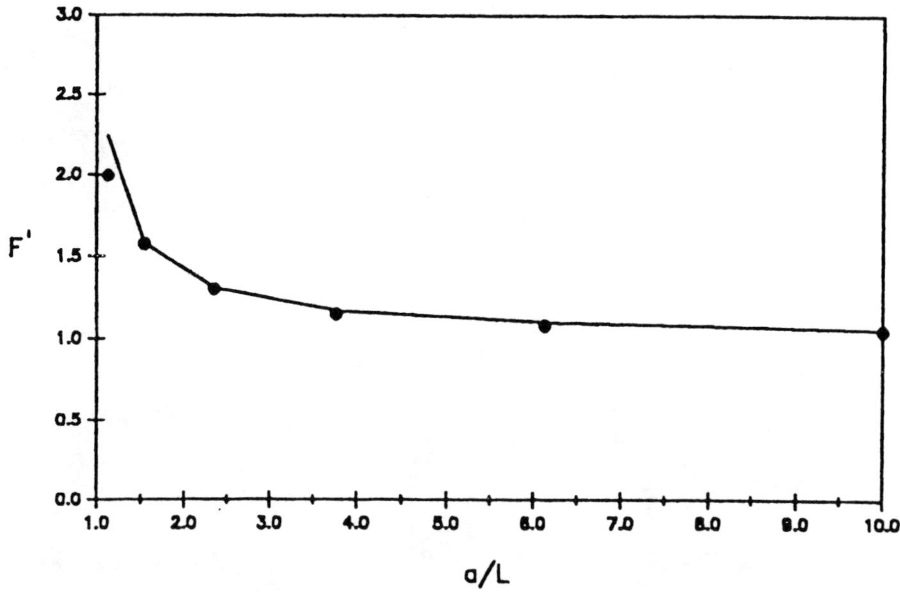

Figure 1. Drag force for the case of a sphere moving parallel to a plane rigid wall, ——: exact solution ; • • ••: integral equation solution.

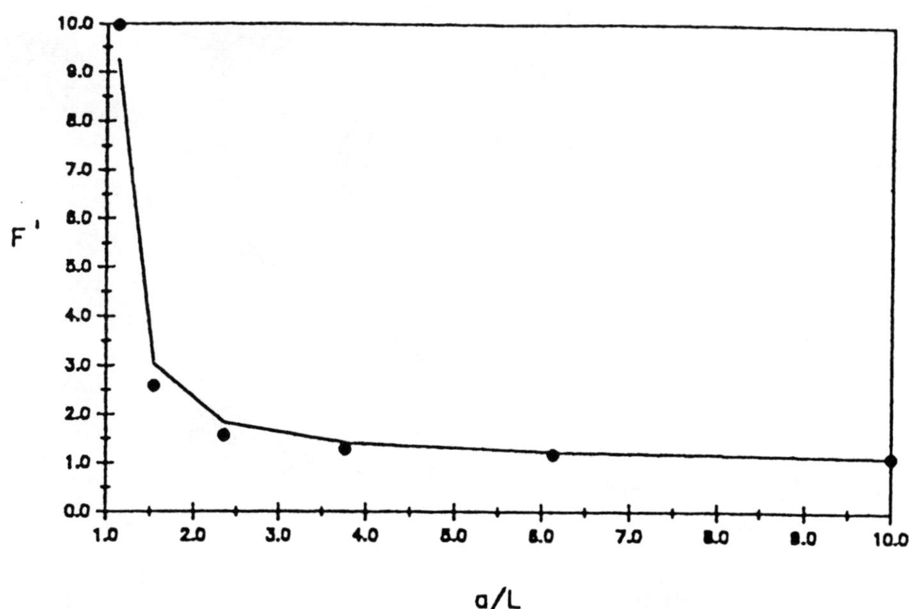

Figure 2. Drag force for the case of a sphere moving perpendicular to a plane rigid wall, ——: exact solution ; • • ••: integral equation solution.

Vorticity-Velocity Formulation for Turbulent Flow by BEM

Z. Rek, L. Skerget, A. Alujevic
University of Maribor, Faculty of Engineering, Smetanova 17, 62000 Maribor, Slovenia

Abstract

This paper deals with the numerical simulation of the turbulent flow using the vorticity-velocity formulation by Boundary Element Method (BEM). A time averaged form of the Navier-Stokes equations is employed through the Reynolds decomposition of the instantaneous value of each variable. Turbulent stress terms are interpreted in the Boussinesq manner and Prandtl's mixing length hypothesis is used. Only algebraic turbulent model is considered in this paper.

Governing equations

Motion of the incompressible viscous fluid is governed by conservation laws of mass and momentum

$$\nabla \cdot \vec{v} = 0 \tag{1}$$

$$(\vec{v} \cdot \nabla)\vec{v} = -\frac{1}{\rho}\nabla p + \nabla \cdot (2\nu \underline{D}) \tag{2}$$

where $\underline{D} = \frac{1}{2}(\nabla \otimes \vec{v} + (\nabla \otimes \vec{v})^T)$ is tensor of the velocity deformations. When considering the turbulent flow a time averaged form of the Navier-Stokes equations is usually employed through the Reynolds decomposition of the instantaneous value of each variable into a time-averaged mean value ($^-$) and an instantaneous fluctuation from the mean value ($'$)

$$\vec{v} = \overline{\vec{v}} + \vec{v'} \tag{3}$$

$$p = \overline{p} + p' \tag{4}$$

Time-averaged Navier-Stokes equations are

$$\nabla \cdot \overline{\vec{v}} = 0 \tag{5}$$

$$(\overline{\vec{v}} \cdot \nabla)\overline{\vec{v}} = -\frac{1}{\rho}\nabla \overline{p} + \nabla \cdot (2\nu \underline{D} - \overline{\vec{v'} \otimes \vec{v'}}) \tag{6}$$

Tensor $-\rho\overline{\vec{v'} \otimes \vec{v'}}$ is Reynolds stress tensor which represents the correlations between the fluctuations of the velocity components. Reynolds stresses are written similarly like laminar ones, known as Boussinesq hypothesis

$$-\rho\overline{\vec{v'} \otimes \vec{v'}} = \rho\nu_t \underline{D} - \frac{2}{3}k\underline{I} \tag{7}$$

by introducing turbulent kinematic viscosity, which is unknown function of space and velocity and must be modelled. The term with k is added to avoid zero normal turbulent stresses and represents turbulent kinetic energy.

Equations which describe turbulent motion of the viscous incompressible fluid are

$$\nabla \cdot \vec{v} = 0 \tag{8}$$

$$(\vec{v} \cdot \nabla)\vec{v} = -\frac{1}{\rho}\nabla p^* + \nabla \cdot \left((2\nu + \nu_t)\underline{D}\right) \tag{9}$$

Label $^-$ is being dropped and all variables are time-averaged, while $p^* = p + \frac{2}{3}\rho k$

Vorticity-velocity formulation

Let's introduce time-averaged vorticity vector

$$\vec{w} = \nabla \times \vec{v} = \left(\frac{\partial v_z}{\partial y} - \frac{\partial v_y}{\partial z}, \frac{\partial v_x}{\partial z} - \frac{\partial v_z}{\partial x}, \frac{\partial v_y}{\partial x} - \frac{\partial v_x}{\partial y}\right) \tag{10}$$

and eq. (9) can be written in the form

$$\vec{w} \times \vec{v} = -\frac{1}{\rho}\nabla P + \nabla \cdot ((2\nu + \nu_t)\underline{D}) \tag{11}$$

where $P = p^* + \rho v^2/2$.

If the curl is taken on both sides and using some vector identities, the next equation is obtained

$$(\vec{v} \cdot \nabla)\vec{w} - (\vec{w} \cdot \nabla)\vec{v} = \nabla \cdot (2\nu\nabla \times \underline{D}) + \nabla \cdot (\nu_t \nabla \times \underline{D}) + \nabla \cdot (\nabla \nu_t \times \underline{D}) \tag{12}$$

The evaluation of the term $\nabla \times \underline{D}$ gives

$$\nabla \times \underline{D} = \frac{1}{2}\nabla \vec{w} \tag{13}$$

Transport equation for the time-averaged vorticity for turbulent flow is then

$$(\vec{v} \cdot \nabla)\vec{w} - (\vec{w} \cdot \nabla)\vec{v} = \nu\nabla^2\vec{w} + \nabla \cdot \left(\frac{\nu_t}{2}\nabla \vec{w}\right) + \nabla \cdot \left(\nabla\left(\frac{\nu_t}{2}\right) \times \underline{D}\right) \tag{14}$$

which is simplified for plane flow to

$$(\vec{v} \cdot \nabla)\vec{w} = \nu\nabla^2\vec{w} + \nabla \cdot \left(\frac{\nu_t}{2}\nabla \vec{w}\right) + \nabla \cdot \vec{f} \tag{15}$$

where

$$\vec{f} = \frac{1}{2}\left(\frac{\partial \nu_t}{\partial x}\left(\frac{\partial v_x}{\partial y} + \frac{\partial v_y}{\partial x}\right) - 2\frac{\partial \nu_t}{\partial y}\frac{\partial v_x}{\partial x}, 2\frac{\partial \nu_t}{\partial x}\frac{\partial v_y}{\partial y} - \frac{\partial \nu_t}{\partial y}\left(\frac{\partial v_x}{\partial y} + \frac{\partial v_y}{\partial x}\right), 0\right) \tag{16}$$

Algebraic turbulence model

By the Prandtl mixing length theory, which assumes the analogy to the theory of gas kinetics, the turbulent viscosity is given by a product of the length scale l_m and velocity scale $\hat{U}$

$$\nu_t = l_m \hat{U} = l_m^2 \left| \frac{\partial v_x}{\partial y} \right| \tag{17}$$

In the algebraic turbulence model, also known as "zero equation model", the turbulent kinetic viscosity is determined from an algebraic equation. There are many models and all of them are good for flows where velocity profile is already developed, but they are not good for flows where recirculation appears. The most simple and known algebraic models are

- model of Nikuradse

$$\frac{l_m}{R} = 0.14 - 0.08\left(1 - \frac{y}{R}\right)^2 - 0.06\left(1 - \frac{y}{R}\right)^4 \tag{18}$$

- model of Van Driest

$$l_m = \kappa y \left[1 - \exp\left(-\frac{y\sqrt{\tau_w/\rho}}{A\nu}\right)\right] \tag{19}$$

where R is pipe radius or half channel height, y is distance from the wall, τ_w is wall shear stress while constants $\kappa = 0.4 and A = 26$.

There is a lot of other algebraic models such as Von Karman's, Cebeci-Smith's, Baldwin-Lomax's, etc. They work quite well if the production and dissipation of the turbulent kinetic energy are balanced. If this is not true then the transport of those quantities must be considered, what implies the usage of the "one" or "two" equation models, such as "$k - \epsilon$" model.

Boundary-domain integral formulation

Using the standard BEM approach the boundary-domain integral equations for kinematics and kinetics are obtained

- kinematics

$$\begin{aligned} c(\xi)v_x(\xi) + \int v_x(s)\frac{\partial u^*(\xi,s)}{\partial n}\,d\Gamma &= \int v_y(s)\frac{\partial u^*(\xi,s)}{\partial t}\,d\Gamma \\ &- \int w(s)\frac{\partial u^*(\xi,s)}{\partial y}\,d\Omega \end{aligned} \tag{20}$$

$$\begin{aligned} c(\xi)v_y(\xi) + \int v_y(s)\frac{\partial u^*(\xi,s)}{\partial n}\,d\Gamma &= -\int v_x(s)\frac{\partial u^*(\xi,s)}{\partial t}\,d\Gamma \\ &+ \int w(s)\frac{\partial u^*(\xi,s)}{\partial x}\,d\Omega \end{aligned} \tag{21}$$

- kinetics

$$\begin{aligned} c(\xi)w(\xi) + \int w(s)\frac{\partial u^*(\xi,s)}{\partial n}\, d\Gamma &= \int \frac{\nu - \nu_t}{\nu}\frac{\partial w(s)}{\partial n} u^*(\xi,s)\, d\Gamma \\ &- \frac{1}{\nu}\int w(s)v_n(s)u^*(\xi,s)\, d\Gamma \\ &+ \frac{1}{\nu}\int \vec{f}(s)\cdot\vec{n}(s)u^*(\xi,s)\, d\Gamma \\ &+ \frac{1}{\nu}\int w(s)\vec{v}(s)\cdot\nabla u^*(\xi,s)\, d\Omega \\ &- \int \frac{\nu_t}{\nu}\nabla w(s)\cdot\nabla u^*(\xi,s)\, d\Omega \\ &- \frac{1}{\nu}\int \vec{f}\cdot\nabla u^*(\xi,s)\, d\Omega \end{aligned} \quad (22)$$

Equations are written in a discrete form what means that the system

$$[A]\{X\} = \{b\} \quad (23)$$

is solved for prescribed boundary conditions. The system is nonlinear, so we are using iterative method with relaxation.

As one can see, there are terms which include derivatives of velocity and turbulent viscosity in kinetic equation. It is worth noting that special care must be given when computing these terms. There are two possibilities.

One is to obtain derivatives by deriving shape functions, like in FEM

$$\nabla u = \{\nabla\Phi\}^T\{U\} \quad (24)$$

But this option is not to be recommended, since it is known that numerical differentiation is an ill conditioned problem.

The other way is to find the gradient of the kinematic equations. Doing so, the derivatives have the same accuracy as the variables, but there is additional work to be done for computation of the required integrals. Derivatives of the velocity components in the domain are obtained from equations

$$\begin{aligned} \frac{\partial v_x(\xi)}{\partial x} &= -\int v_x(s)\frac{\partial}{\partial x_\xi}\left(\frac{\partial u^*(\xi,s)}{\partial n}\right) d\Gamma + \int v_y(s)\frac{\partial}{\partial x_\xi}\left(\frac{\partial u^*(\xi,s)}{\partial t}\right) d\Gamma \\ &- \int w(s)\frac{\partial}{\partial x_\xi}\left(\frac{\partial u^*(\xi,s)}{\partial y}\right) d\Omega \end{aligned} \quad (25)$$

$$\begin{aligned} \frac{\partial v_x(\xi)}{\partial y} &= -\int v_x(s)\frac{\partial}{\partial y_\xi}\left(\frac{\partial u^*(\xi,s)}{\partial n}\right) d\Gamma + \int v_y(s)\frac{\partial}{\partial y_\xi}\left(\frac{\partial u^*(\xi,s)}{\partial t}\right) d\Gamma \\ &- \int w(s)\frac{\partial}{\partial y_\xi}\left(\frac{\partial u^*(\xi,s)}{\partial y}\right) d\Omega \end{aligned} \quad (26)$$

$$
\begin{aligned}
\frac{\partial v_y(\xi)}{\partial x} &= -\int v_y(s)\frac{\partial}{\partial x_\xi}\left(\frac{\partial u^*(\xi,s)}{\partial n}\right)d\Gamma - \int v_x(s)\frac{\partial}{\partial x_\xi}\left(\frac{\partial u^*(\xi,s)}{\partial t}\right)d\Gamma \\
&+ \int w(s)\frac{\partial}{\partial x_\xi}\left(\frac{\partial u^*(\xi,s)}{\partial x}\right)d\Omega \qquad (27) \\
\frac{\partial v_y(\xi)}{\partial y} &= -\int v_y(s)\frac{\partial}{\partial y_\xi}\left(\frac{\partial u^*(\xi,s)}{\partial n}\right)d\Gamma - \int v_x(s)\frac{\partial}{\partial y_\xi}\left(\frac{\partial u^*(\xi,s)}{\partial t}\right)d\Gamma \\
&+ \int w(s)\frac{\partial}{\partial y_\xi}\left(\frac{\partial u^*(\xi,s)}{\partial x}\right)d\Omega \qquad (28)
\end{aligned}
$$

Test case – turbulent flow in the channel

As a test case, the developed turbulent velocity profile in the channel is considered with vorticity-velocity BEM formulation for Re=2500. Turbulent viscosity is modelled by the algebraic model

$$\nu_t = l_m^2 \left|\frac{\partial U}{\partial y}\right| \qquad (29)$$

where l_m is Prandtl's mixing length. Nikuradse

$$l_m/R = 0.14 - 0.08\left(1-\frac{y'}{R}\right)^2 - 0.06\left(1-\frac{y'}{R}\right)^4$$

and Van Driest

$$l_m = \kappa y'\left[1-\exp\left(-\frac{y'\sqrt{\tau_w/\rho}}{A\nu}\right)\right]$$

equations for mixing length have been used. In the beginning of the channel the uniform velocity is prescribed, while at the end, the zero vorticity flux is assumed. Geometry data and boundary conditions are shown on the figure 1.

The channel is discretized with 40 boundary elements and 100 internal cells (10 × 10), totalling 441 nodes. In the y direction, the mesh is condensed near the wall, and the ratio between the smallest and the largest element is 1:4, so the boundary layer is better described. Quadratic interpolation functions are used. Discrete model is shown on the Figure 2.

Figure 3 shows development of the turbulent velocity profile for Van Driest model along the channel length from uniform at the beginning to the fully developed at the end.

Figure 4 shows the comparation of the velocity profiles at the distance $x = 36H$ between both algebraic models and results of the FEM analysis model ($k-\epsilon$ with wall functions).

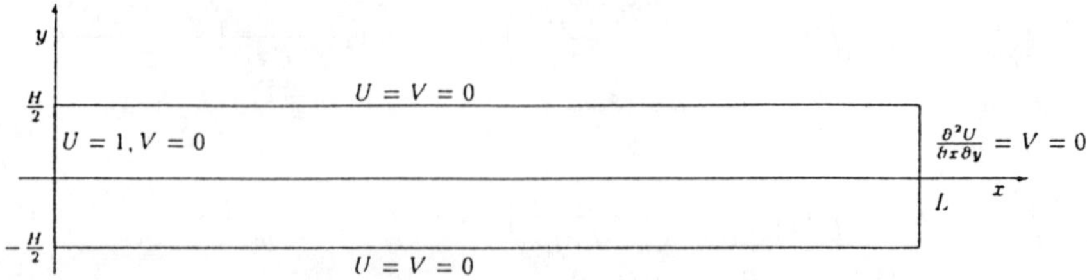

Figure 1: Geometry of the channel and boundary conditions, $H = 1, L = 40$.

Conclusion

The vorticity-velocity formulation of the BEM analysis has proved to be very successful also in the case of turbulent flow. Various models for the turbulent viscosity modelling can be used. In this paper, only the algebraic models are considered, but no special difficulty arises if well known $k - \epsilon$ model is included, except the neediness of a "big computer".

It has been also mentioned that special care has to be taken when computation of velocity gradients is performed. For points in the domain, there is no problem. From ordinary boundary-domain integral equations only the normal derivatives can be obtained for boundary nodes. If a gradient has to be known, the only way leads to the hypersingular equations. Tests made on the potential problems shows a great advantage of the hypersingular formulation against ordinary strongly singular method. The future work will be performed in this direction.

References

[1] Alujevic A., Kuhn G., Skerget P.: "Boundary Elements for the Solution of Navier-Stokes Equations", **Computer Methods in Applied Mechanics and Engineering**, p.p. 1187-1201,, 1991.

[2] Brebbia C.A., Telles J.F.C., Wrobel L.C.: **Boundary Element Methods — Theory and Applications**, Springer-Verlag, New York, 1984.

[3] Chien K.Y.: "Predictions of Channel and Boundary-Layer Flows with a Low Reynolds Number Turbulence Model", **AIAA Journal**, Vol. 20, No.1, 1982.

[4] Martinuzzi R., Pollard A.: "Comparative Study of Turbulence Models in Predicting Turbulent Pipe Flow, Part I: Algebraic Stress and $k-\epsilon$ Models", **AIAA Journal**, Vol. 27, p.p. 29-36, No. 1, 1989.

[5] Nagano Y., Kim C.: "A Two-Equation Model for Heat Transport in Wall Turbulent Shear Flows", **Journal of Heat Transfer**, Vol. 110, p.p. 583-589, 1987.

[6] Patel V.C., Rodi W., Scheurer G.: "Evaluation of Turbulence Models for Near-Wall and Low-Reynolds Number Flows", **Proceedings, 3rd Symposium on Turbulent Shear Flows, University of California**, 1981.

[7] Ruprecht A.: "Turbulence modeling", *CFD'90 - Intensive Course on Computational Fluid Mechanics and Heat Transfer*, Turboistitut, Ljubljana, 1990.

[8] Skerget P., Kuhn G., Alujevic A., Brebbia C.A.: "Time Depended Transport Problems by BEM", **Advances in Water Resources**, Vol. 12, p.p. 9-20, No. 1, 1989.

[9] Tennekes H., Lumly J.L.: **A first Course in Turbulence**, Boston, The MIT Press, 1972.

[10] Tosaka N., Kakuda K.: "Numerical Simulations of Laminar and Turbulent Flows by Using an Integral Equations", **BEM IX**, Vol. 3, 1987.

[11] Tong G.D.: "Fundamental considerations in Computational Fluid Mechanics", **Eight Australasian Fluid Mechanics Conference**, University of Newcastle, N.S.W., 1983.

Figure 2: Discrete model.

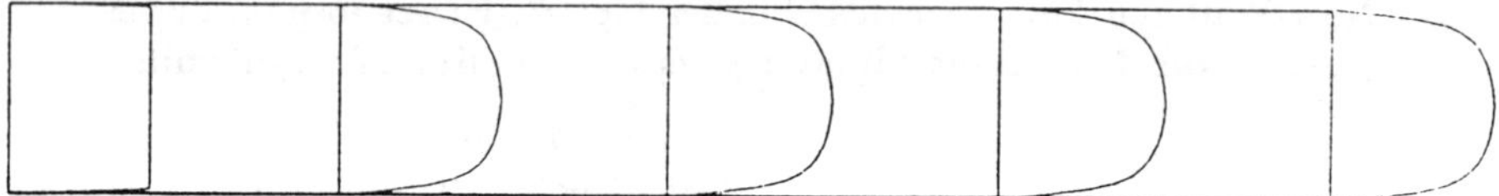

Figure 3: Velocity profiles U along the channel.

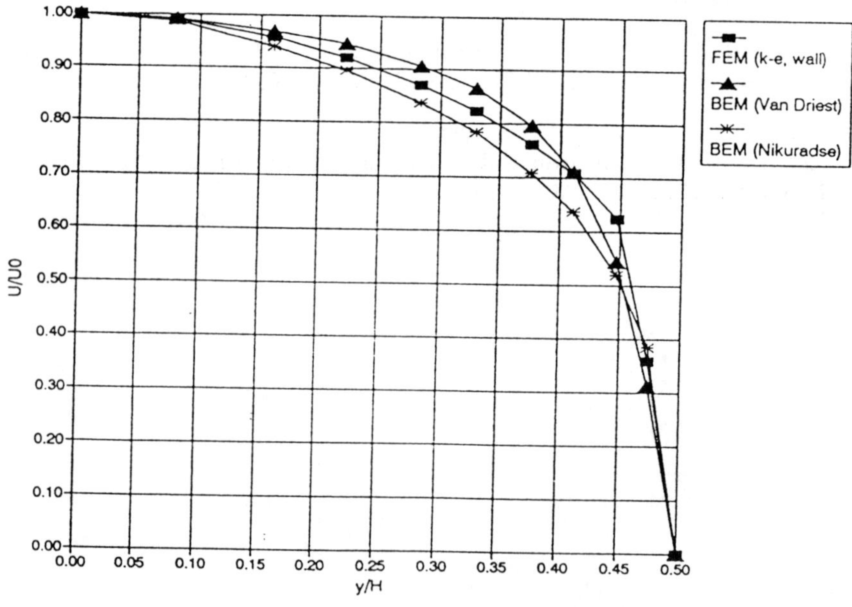

Figure 4: Comparison of the velocity profiles at $x = 36H$.

Including Sharp Edge Vortex Shedding in Boundary Integral Solutions for Two Dimensional Potential Flows

L.H. Wong, S. Calisal

Department of Mechanical Engineering, University of British Columbia, 2324 Main Mall, Vancouver, B.C., V6T 1Z4, Canada

ABSTRACT

A discrete vortex method was incorporated into a time domain boundary element algorithm for the numerical simulation of normal oscillating flow past a flat plate. Significant computational advantages result because of the relatively simple approach to the handling of separation at the sharp edges while working only with the boundary values.

The separated vortex sheet issuing from a sharp edge in normal flow is modelled by a series of discrete vortices introduced one at a time into the flow field at given time intervals. The motion of each vortex is traced over time using its convection velocity. For low Keulegan-Carpenter numbers, vortex shedding takes place close to the edge. The discrete vortex method can, in such cases, be looked upon as the inner region solution to the problem of flow past the normal plate. This inner region solution has to be matched with the outer potential flow solution. The combination of boundary element and discrete vortex methods provides this matching and at the same time do not require calculations inside the domain.

INTRODUCTION

There are numerous instances in the study of fluid dynamics where vortex shedding forces are significant and should be taken into account. A good example is in the numerical prediction of the roll response of a ship with sharp corners and bilge keels where wave damping is light and vortex effects are important.

Flow separation usually occurs when the boundary layer on a body surface reaches a sharp edge where the radius of curvature of the edge is very much smaller than the boundary layer thickness. In two dimensional flow, the adverse pressure gradient set up results in the formation of a shear layer which subsequently rolls up into a tight spiral and is then shed into the flow field as a free vortex sheet of infinitessimal thickness. Such a vortex sheet can be approximated by the introduction of discrete vortices at given time intervals as the flow develops. A persistent and yet unsolved problem in such schemes is the irregular roll-up of the vortex sheet as well as numerical instability due to the uncharacteristically high velocities induced when a vortex is very near to another vortex or its image. These problems have been the focus of study in research on vortex methods since the early 70s.

Chorin [1] used vortices with a viscous core to model viscous diffusion as well as to stabilize his numerical procedure. As a result, the maximum induced velocity of a vortex is finite at a given distance from the vortex position and decreases to zero as the centre of the vortex is approached. Clements and Maull [2], to obtain stable solutions, limited the induced velocities by amalgamating any pair of vortices that are too close together. Fink and Soh [3] pioneered a scheme of rediscretization in which the vortex sheet is rearranged into equidistant positions after each time step in the numerical procedure. The above and many other similar schemes help in one way or another to give smooth vortex roll-up as well as extend the computational time span in which calculations remain stable, Wong [4].

Graham [5] applied the discrete vortex method to calculate the vortex forces induced at a sharp edge in oscillatory motion at low Keulegan-Carpenter numbers. This was done by regarding vortex shedding from an infinite wedge as the inner region of flow past a large but finite body. The underlying assumption in this case is that the body length scale is large so that the vortex shed does not affect other parts of the body which are far away from the edge when compared to the flow length scale. Downie, Bearman and Graham [6] followed up on this and calculated the vortex damping forces on a rolling barge.

Like all other computationally intensive schemes, the boundary element method has undergone significant development in the last two decades with the advent of high speed and large memory computers. An advantage of this method is that for potential flow calculations only conditions on the boundary need to be specified and calculations are carried out along the discretized boundary only. This method is well suited for application in boundary value potential flow simulations involving relatively complex and possibly time varying geometries. It is realized that BEM is compatible with the discrete vortex model in that each discrete vortex can be treated as an internal singularity which can be handled using an analogue of the residue theorem in complex analysis.

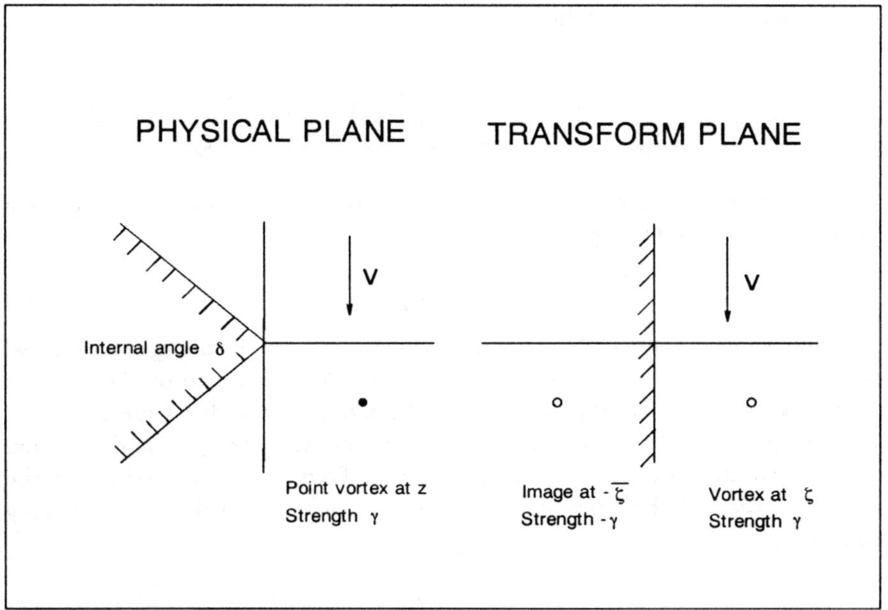

Figure 1: The physical Z-plane and the transformed ζ-plane.

DISCRETE VORTEX METHOD

The discrete vortex method (DVM) is a time stepping procedure [4] which models the shear layer issuing from a sharp edge using discrete vortices. In the formulation, an infinite sharp wedge is mapped onto a half plane via a Schwarz-Christoffel transformation, Fig 1, using the following equation:

$$Z = \frac{M}{\lambda}\zeta^{\lambda} \tag{1}$$

where M is a scaling constant between the physical z and the transform ζ planes and $\lambda = 2 - \delta/\pi$ is a parameter dependent on the internal angle δ of the wedge.

The non-dimensional external flow velocity, $v = \sin 2\pi\tau$, where $\tau = t/T$ (T being the period of oscillation) is in a direction normal to the wedge bisector.The strength of the vortex at z and its corresponding point ζ is given by γ, Fig 1. The complex potential in the ζ-plane, with n vortices in the field, is given by:

$$F(\zeta) = iV\zeta + \frac{i}{2\pi}\sum_{k=0}^{n}\Gamma_k \ln\frac{\zeta - \zeta_k}{\zeta + \bar{\zeta}_k} \tag{2}$$

Applying Routh's correction and normalising all variables (after Graham [5]) result in an expression for the complex conjugate velocity at a point j

in the physical plane as follows:

$$w_j = \frac{i\zeta_j}{\lambda z_j}[v + \sum_{k=0,k\neq j}^{n} \frac{\gamma_k}{2\pi}(\frac{1}{\zeta_j - \zeta_k} - \frac{1}{\zeta_j + \bar{\zeta}_k}) - \frac{\gamma_j}{2\pi}(\frac{1}{\zeta_j + \bar{\zeta}_j} - \frac{1-\lambda}{2\zeta_j})] \quad (3)$$

Assuming that all vortices convect with the flow, the convection equation in the physical plane is given by:

$$\dot{z}_k = \bar{w}_k \quad (4)$$

A new vortex (called the nascent vortex) is introduced along the external wedge bisector at the end of each time step. The distance of the nascent vortex from the sharp edge is taken to be a function of the time step size, the wedge angle, the free stream velocity and the strength and location of all the other vortices already in the flow field. The expression for the initial location of the nascent vortex (z_0) can be found using the following equations:

$$z_0 = (K\Delta\tau)^m \quad (5)$$

where

$$\begin{aligned} m &= \frac{\lambda}{2\lambda - 1} \\ K &= \frac{v_n\sqrt{4-\lambda}(\lambda-1)(2\lambda-1)}{2\pi\lambda^3} \end{aligned} \quad (6)$$

with v_n defined in Equation (9).

The strength of the nascent vortex can then be obtained using the Kutta condition as follows:

$$\begin{aligned} \frac{dF(\zeta)}{d\zeta}\bigg|_{\zeta=0} &= i[v - \sum_{k=0}^{n} \frac{\gamma_k}{2\pi}(\frac{1}{\zeta_k} + \frac{1}{\bar{\zeta}_k})] \\ &= 0 \end{aligned} \quad (7)$$

$$\gamma_0 = v_n \frac{|\zeta_0|^2}{Re(\zeta_0)} \quad (8)$$

where

$$v_n = \pi sin(2\pi\tau) - \sum_{k=1}^{n} \gamma_k \frac{Re(\zeta_k)}{|\zeta_k|^2} \quad (9)$$

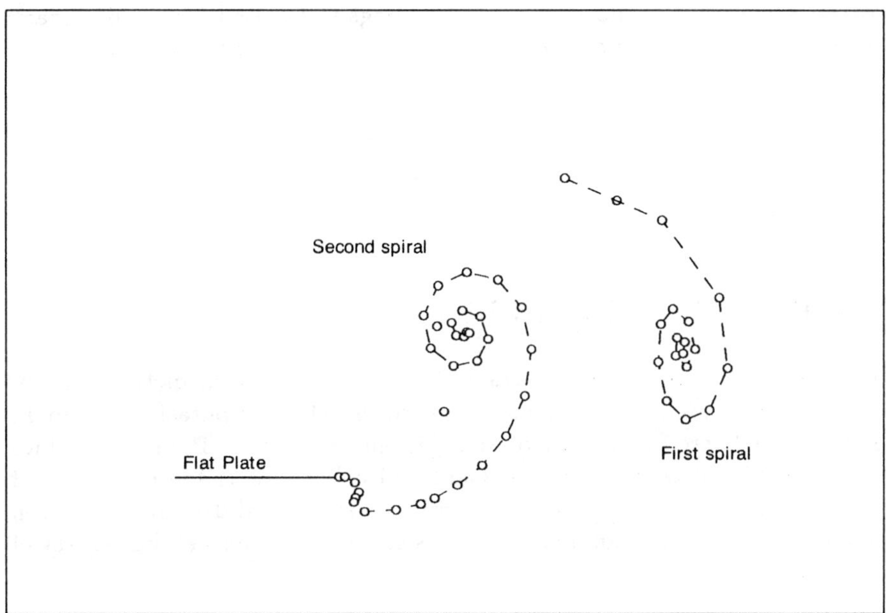

Figure 2: Vortex sheet rollup at $\tau = 1.00$ for a flat plate.

A vortex decay mechanism is used to account for the vortex pairing process as described in [5] and provide a model for vortex diffusion. For this a simple exponential rule reduces the effective strength of each vortex to one half of its original value at the end of one time cycle, i.e.

$$\gamma(t) = \gamma_0[1 - \exp(-0.6932/\tau)] \tag{10}$$

Although violating conservation of vorticity, this vortex decay model is successful in maintaining computational stability over a large number of flow cycles. Also, reducing circulation in the wake limits the magnitude of the vorticity introduced. Consequently, the otherwise over-predicted vortex induced forces are kept at levels which are comparable to experimentally determined values. Vortex induced forces can be calculated using the unsteady form of the Blasius force equation. However, in this work, vortex forces need not be explicitly calculated since the discrete vortex method is to be used in a boundary element scheme. An example of the calculated vortex roll-up for a flat plate at the end of a cycle is given in Fig 2.

The simulation of vortex shedding from a finite wedge oscillating in unbounded fluid requires some length scale modifications to the above calculations. In other words, the vortex strengths and positions calculated have to be adjusted to account for changes in Keulegan-Carpenter numbers (K_c). Graham [5] showed that the ratio of the length scales of the infinite and finite wedges is proportional to $(K_c)^{\frac{\lambda}{2\lambda-1}}$. The resultant non-dimensional vortex strength (γ_B) and length (L_B) scalings to be used for the boundary element procedure for finite K_c and body length ($2d$) is then given by:

$$\begin{aligned} \gamma_B &= \gamma(2K_c)^{\frac{1}{2\lambda-1}} \\ Z_B &= z(2K_c)^{\frac{\lambda}{2\lambda-1}} \end{aligned} \tag{11}$$

VORTEX SHEDDING IN BEM

In the present application, the standard boundary element method for potential flow is used. The boundary enclosing the computational domain is discretized into N number of straight line elements. Boundary values are assumed constant on each element and collocated at the mid-point of the element. Either the velocity potential or its normal derivative at given nodal points on the boundary emerge as solutions to the well-known set of equations given below.

$$\sum_{j=1}^{N} \phi_j(\frac{\partial G}{\partial n})_{ij} - \sum_{j=1}^{N} G_{ij}(\frac{\partial \phi}{\partial n})_j = 0 \tag{12}$$

where $i = 1 \ldots N$ and G is the weighting source function taken to be $G = \ln r$, r being the distance between the point i and any other point on the boundary j. In the above, $\frac{\partial}{\partial n}$ is the normal derivative of the given function. The system of N equations in N unknowns are written as a single matrix equation. A four point Gaussian quadrature is used to evaluate the non-diagonal members of G and H while analytical solutions can be used if $i = j$, [11]. The following expressions for G and H apply for the boundary elements:

$$\begin{aligned} G_{ij} &= \frac{S_j}{2}\sum_{q=1}^{4} \omega_q \ln(r) && i \neq j \\ &= S_j[\ln(\frac{S_j}{2}) - 1] && i = j \\ H_{ij} &= -\frac{S_j}{2}\sum_{q=1}^{4} \omega_q \frac{D_n}{r^2} && i \neq j \\ &= -\pi && i = j \end{aligned} \tag{13}$$

where D_n is the perpendicular distance, r is the actual distance from the point i to the Gauss point q on element j, S_j is the length of element j and ω_q is the weight for the Gauss integration.

Since Laplace's equation is linear, the total potential at any boundary node j can be taken to be:

$$\phi = \phi_1 + \phi_v \tag{14}$$

where ϕ_1 and ϕ_v are the potentials due to the potential flow and the shed point vortices respectively. The same applies for the normal derivative of the velocity potential, that is,

$$\frac{\partial \phi}{\partial n} = (\frac{\partial \phi}{\partial n})_1 + (\frac{\partial \phi}{\partial n})_v \tag{15}$$

In time domain simulations, the boundary values and vortex strengths are to be prescribed for the next time step $(t + \delta t)$ in order to march forward in time. The matrices G and H need to be calculated only once for the rigid sharp edged section. The strengths and positions of all discrete vortices in the flow at time $(t + \delta t)$ are found using DVM. The potential and velocity induced by the M number of discrete vortices in the flow on any boundary node j is given by:

$$\phi_{v_j} = \sum_{k=1}^{M} \frac{\gamma_k \theta_{jk}}{2\pi}$$

$$(\frac{\partial \phi}{\partial n})_{v_j} = - \sum_{k=1}^{M} \frac{\gamma_k}{2\pi R_{jk}}$$

where θ_{jk} is the angular position of the node j with reference to the kth vortex with strength γ_k and at a distance R_{jk} from the node j. For this application, all the vorticity shed from a particular edge are placed at a point very near to that edge. In other words, we 'average' all the vortex positons so that the influence of all discrete vortices on any given boundary element is calculated through the relative positions of that node (j) and a fixed point near the edge. This simplification is acceptable for small K_c since much of the vortex shedding effects are confined to a region near to the edge. Consequently, if we include the vortex induced normal velocities on the body, Equation (13) can be rewritten in the following discretized form:

$$\sum_{j=1}^{N} \phi_j H_{ij} = \sum_{j=1}^{N} G_{ij} [(\frac{\partial \phi}{\partial n})_j - (\frac{\partial \phi}{\partial n})_{v_j}] - \phi_{v_j} H_{ij} \tag{16}$$

for $i = 1 \ldots N$. This system of N equations are then solved using Gaussian elimination for the unknowns ϕ_j.

FORCES ON A FLAT PLATE

Although this external flow problem can be solved using DVM by itself, this example serves to illustrate the time domain implementation of the procedure given in the last section. Published numerical and experimental results, for example those in [5], [12] and [13] could be used for comparison purposes: not to judge the relative merits of the numerical schemes proposed by other researchers but as a test case for the present method. On the body, with the internal angle at the wedge being δ, the non-dimensional normal velocity is given by:

$$\frac{\partial \phi}{\partial n} = \sin(2\pi\tau)\cos(\delta/2) \tag{17}$$

where the non-dimensional elapsed time is $\tau = t/T$, T being the period of oscillation of the wedge. In the limit $\delta \to 0$, the problem becomes one of a normal oscillating flat plate. The dimensional hydrodynamic force per unit depth acting on S_b at any given instant is taken to be:

$$F = \oint_{S_b} p.\vec{n}\, dS_b$$

where p is the hydrodynamic pressure acting on the discrete element dS_b with normal vector $\vec{n}$. The calculated value of F is then normalized using $\frac{1}{2}\rho V_0^2 d$ to give a non-dimensional force coefficient C_F where:

$$C_F = \frac{2\pi}{K_c} F_B \tag{18}$$

In the above, F_B is the non-dimensional force on the plate such that $F = (4\pi\rho\frac{V_0}{T}d^2)F_B$. Values of C_F for a flat plate oscillating at $K_c = 6.6$ are plotted together with experimental results from [14] in Fig 3. The present method predicted C_F values that are in excellent agreement with experimental results.

Drag (C_d) and inertia (C_a) coefficients are found by taking Fourier integrals of Morison's equation over a time cycle and given thus:

$$C_d = \frac{3\pi}{4}\int_{n\tau}^{(n+1)\tau} C_F \sin(2\pi\tau)d\tau \tag{19}$$

$$C_a = \frac{2K_c}{\pi^2}\int_{n\tau}^{(n+1)\tau} C_F \cos(2\pi\tau)d\tau \tag{20}$$

(21)

Calculated drag (C_d) and inertia (C_a) coefficients are presented in Fig 4 and Fig 5. Predicted C_d for the flat plate agree well with experimental and numerical results from [12] and [13]. At K_c above 3.0, predicted C_d appears to underestimate experimental values. However, Obasaju's unpublished data (as reported in [5]) showed C_d values lower than those found using the present method. Therefore, given the variation of prevailing conditions under which different researchers obtained their experimental results, the present method predicts drag forces on the normal flat plate very well.

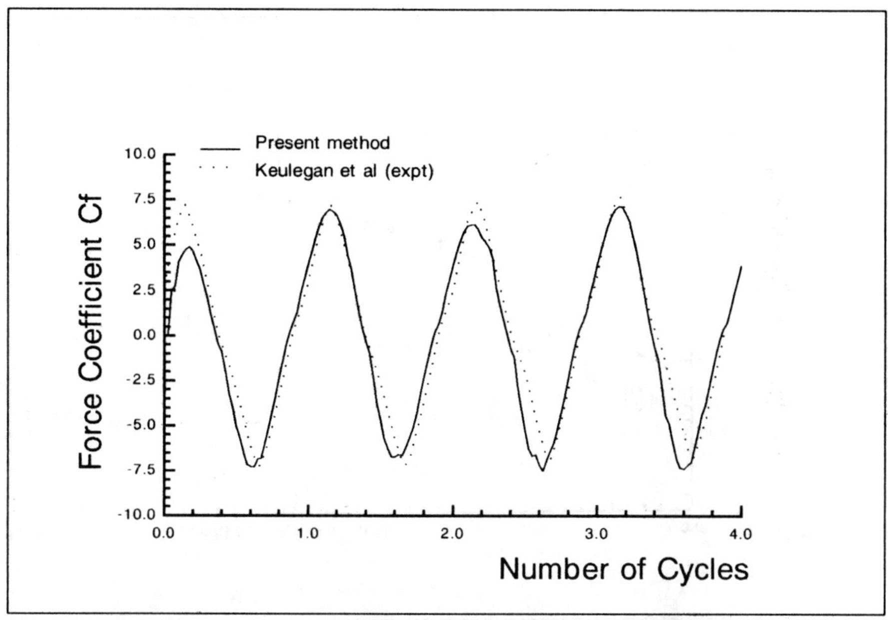

Figure 3: Forces on an oscillating flat plate, $K_c = 6.6$.

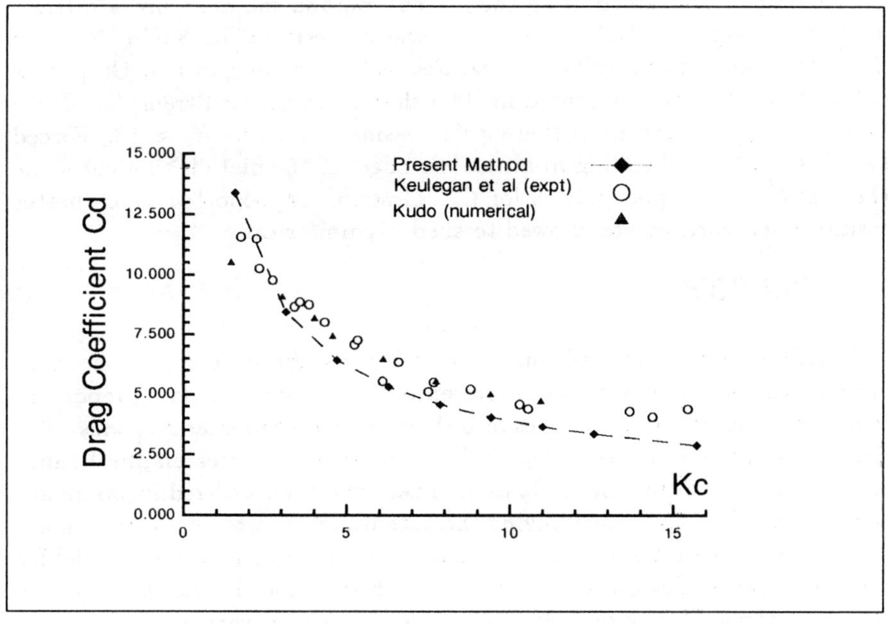

Figure 4: Drag coefficients for the flat plate.

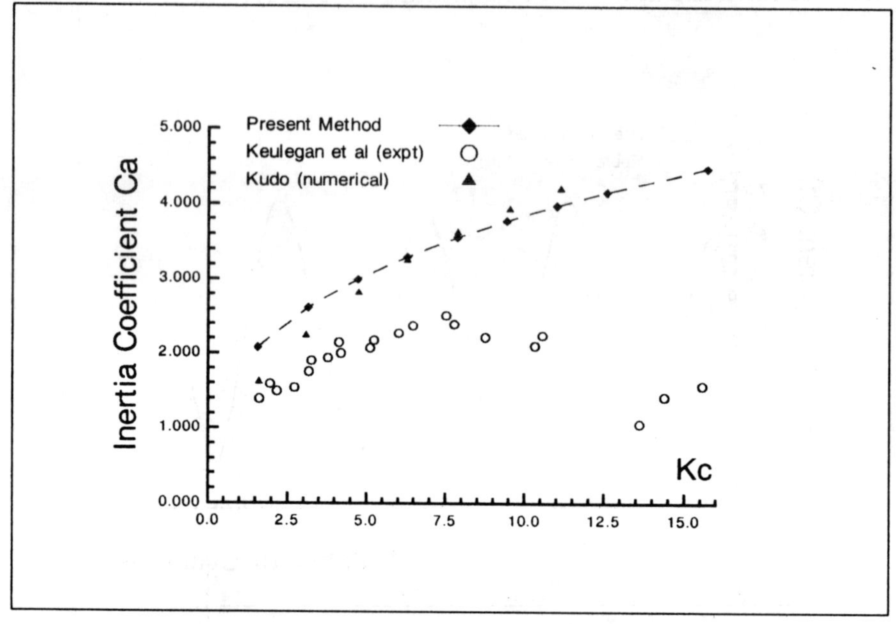

Figure 5: Inertia coefficients for the flat plate.

Values of C_a obtained using the present method shows an upward trend with increasing K_c which is similar to those presented by Kudo [12]. On the other hand the experimental results of both Keulegan and Carpenter [13] and Obasaju (as presented in [14]) showed an upward trend for C_a up to about $K_c \approx 7$ and from there a decreasing C_a, up to $K_c \approx 18$. Forced symmetric vortex shedding from the two edges of the plate is thought to be the reason for the poor match for C_a. Graham [14] reported much better results when vortices are allowed to shed asymmetrically.

CONCLUSION

A domain zoning procedure which accounts for sharp edge vortex shedding in two dimensional flow has been presented. The vortex shedding model introduces potential point vortices into the flow field at regular intervals. The potentials and normal velocities induced by these vortices on the boundary of the domain can be evaluated. These are then utilized in boundary element calculations, hence solving the potential flow problem very economically. Use of the same transform equation in the discrete vortex model for different edge angles provide a useful and flexible tool in the investigation of a wide range of flow problems involving sharp edge vortex shedding.

The present method has been successfully implemented for the calculation of drag and inertia forces on oscillating flat plate. Results obtained for the flat plate agreed quite well with published experimental and numerical results of other researchers. This method can also be used for oscillating square edged cylinders without modification.

Although the discrete vortex model can be applied in conjunction with other numerical methods for the given class of problems, it is uncertain whether such an exercise would be more fruitful than working directly with viscous flow methods since the advantage of working with less elements (i.e. elements on the boundary only) is lost.

ACKNOWLEDGEMENT

The work described in this paper was made possible through financial assistance from the Natural Sciences and Engineering Research Council of Canada (NSERC).

References

[1] Chorin A.J., 'Numerical study of slightly viscous flow', J Fluid Mech, Vol 57, pp 787, 1973.

[2] Clements R.R. and Maull D.J., 'The representation of sheets of vorticity by discrete vortices', Prog Aerospace Science, Vol 16(2), pp 129, 1975.

[3] Fink P.T. and Soh W.K., 'Calculation of vortex sheets in unsteady flow and applications in ship hydrodynamics', Proceedings of the 10th Symp on Naval Hydrodynamics, pp463-491, Cambridge, Massachussetts, 1974.

[4] Wong L.H. and Calisal S.M., 'Vortex shedding from sharp edges in oscillatory flow', Proceedings of the 2nd World Conf on Expt Heat Transfer, Fluid Mech and Thermodynamics, pp 232-239, Dubrovnik, Yugoslavia, 1991.

[5] Graham J.M.R., 'The forces on sharp-edged cylinders in oscillatory flow at low Keulegan-Carpenter numbers', J Fluid Mech, Vol 97(I), pp 331-346, 1980.

[6] Downie M.J., Bearman P.W. and Graham J.M.R., 'Effect of vortex shedding on the coupled roll response of bodies in waves', J Fluid Mech, Vol 189, pp 243-264, 1988.

[7] Faltinsen O.M., 'A numerical nonlinear method of sloshing in tanks with two dimensional flow', J Ship Res, Vol 22(3), pp 193-202, 1978.

[8] Grosenbaugh M.A. and Yeung R.W., 'Nonlinear free-surface flow at a two-dimensional bow', J Fluid Mech, Vol 209, pp 57-75, 1989.

[9] Isaacson M. and Cheung K.F., 'Second order wave diffraction around two-dimensional bodies by time domain method', Applied Ocean Res, Vol 13(3), 1991.

[10] Pawlowski J.S., Baddour R.E. and Hookey N.A., 'The development of second generation direct solution numerical models of large motions of floating bodies in waves', Proceedings of Marine Dynamics Conference, pp 43-49, St John's, Newfoundland, 1991.

[11] Brebbia C.A., The Boundary Element Method for Engineers, Pentech Press, London, 1978.

[12] Kudo K., 'Hydrodynamic forces of the oscillating flat plate, Proc 3rd Int Conf Num Ship Hydrodynamics', pp 347-357, Paris, 1981.

[13] Keulegan G.H. and Carpenter L.H., 'Forces on cylinders and plates in an oscillating fluid', J R National Bureau of Standards, Vol 60(5), 1958.

[14] Graham J.M.R., 'Numerical simulation of steady and unsteady flow about sharp-edged bodies', Proc Int Symp on Separated Flow Around Marine Structures, pp 347-373, Trondheim, 1985.

A Boundary Element Method for Steady-State Two-Dimensional Stokes Flows and Its Asymptotic Error Estimates

H. Wang
Department of Mathematics, University of Wyoming, Laramie, WY 82071, U.S.A.

ABSTRACT

In this paper we present a boundary element method to solve steady-state two-dimensional Stokes flows. A numerical scheme is developed. An optimal-order error estimate in energy norm and superconvergence results in L^∞ norm are obtained.

INTRODUCTION

It is well-known that Navier-Stokes equations are very important in various applications. Numerically, we can solve Navier-Stokes equations by solving linear Stokes equations iteratively. The study of Stokes problem itself also provides a basis for the study of Navier-Stokes problem.

For Stokes problem, both "stream function" form and "velocity-pressure" form are available, which correspond to a biharmonic or a saddle-point problem, respectively. If we use domain-type methods to solve Stokes problems, we need to solve either a fourth-order problem or a saddle-point problem. For the saddle-point problem, the incompressibility condition "$\nabla \cdot \mathbf{u} = 0$" requires a balance between the finite element spaces for the velocity and for the pressure in the form of the "inf-sup" condition. For a general domain, some finite element spaces based on triangular partition may not be optimal in terms of the polynomials these finite element spaces use and the convergence rate these spaces possess, for example see Girault [3].

The second class of numerical methods are boundary element methods (BEM). BEM reduce the dimension of the problems by one and constitute a very convenient manner of treating unbounded regions. In this paper, we present a BEM scheme to solve steady-state two-dimensional Stokes flows. The solution of the problem is expressed in terms of a simple layer potential that is valid for both interior and exterior problems simultaneously. Moreover, the incompressibility constraint "$\nabla \cdot \mathbf{u} = 0$" is *automatically* satisfied. The problems of finding the velocity $\mathbf{u}$ and the pressure p are separated. Numerical results will be presented on the conference. In this paper, we obtain an optimal-order error estimate in energy norm and superconvergence results in L^∞ norm. Furthermore, any order derivatives of the numerical solution have same-order superconvergence outside a neighbourhood of the boundary.

VARIATIONAL PRINCIPLE

In this paper, we solve the Dirichlet problem of steady-state two-dimensional Stokes equations:

$$\begin{cases} -\nu\Delta\mathbf{u} + \nabla p &= \mathbf{f}(\mathbf{x}), \qquad \mathbf{x} \in \Omega \cup \Omega', \\ \nabla \cdot \mathbf{u}(\mathbf{x}) &= 0, \qquad \mathbf{x} \in \Omega \cup \Omega', \\ \mathbf{u}(\mathbf{x}) &= \mathbf{u}_0(\mathbf{x}), \qquad \mathbf{x} \in S, \end{cases} \tag{1}$$

where Ω is a simply-connected domain with a piecewise-smooth boundary $S = \partial\Omega$, $\Omega' = R^2 - \bar{\Omega}$. The unknown variables are fluid rate $\mathbf{u} = (u_1, u_2)$ and pressure p of viscous incompressible fluid flows that fill Ω or Ω'. ν is the kinematic viscosity. $\mathbf{u}_0$ and $\mathbf{f}$ are given functions.

A special solution $\mathbf{u}^{\mathbf{f}}$ for the first two equations in problem (1) can be obtained analytically in terms of a volume potential, see Ladyzhenskaya [4]. Many papers have discussed the efficient computations for $\mathbf{u}^{\mathbf{f}}$, we do not discuss this issue further here. Then, $\mathbf{u}^{**} = \mathbf{u} - \mathbf{u}^{\mathbf{f}}$ satisfies problem (1) with a homogeneous right-hand side. Thus, we only consider problem (1) with $\mathbf{f} \equiv \mathbf{0}$ in this paper.

In this paper, we use the following Sobolev spaces defined in Ladyzhenskaya [4], Lions and Magenes [5], Nedelec [6]:

$$
\begin{aligned}
H^1(\Omega) &= \left\{ v(\mathbf{x});\ v \in L^2(\Omega), \nabla v \in \left(L^2(\Omega)\right)^2 \right\}, \\
W_0^1(\Omega') &= \left\{ v(\mathbf{x});\ \frac{v}{\sqrt{1+|\mathbf{x}|^2}\ \ln(2+|\mathbf{x}|^2)} \in L^2(\Omega'), \nabla v \in \left(L^2(\Omega')\right)^2 \right\}, \\
\mathbf{W}(R^2) &= \left\{ \mathbf{v}(\mathbf{x}) \in \left(W_0^1(R^2)\right)^2;\ \ \nabla \cdot \mathbf{v} = 0,\ \ \forall \mathbf{x} \in \Omega \cup \Omega' \right\}, \\
H^r(S) &= \text{the standard Sobolev space defined on } S \text{ with index } r, \\
\mathbf{H}^r &= (H^r(S))^2, \\
\mathbf{H}_0^r &= \left\{ \mathbf{u}_0 \in \mathbf{H}^r;\ \int_S \mathbf{u}_0 \cdot \mathbf{n}\ ds = 0 \right\}, \\
\bar{\mathbf{H}}^r &= \left\{ \mathbf{g} \in \mathbf{H}^r;\ \int_S \mathbf{g} \cdot \mathbf{c}\ ds = 0, \qquad \forall \mathbf{c} = (c_1, c_2) \in R^2 \right\}, \\
\mathbf{H}_E^r &= \mathbf{H}^r/E, \qquad \bar{\mathbf{H}}_E^r = \bar{\mathbf{H}}^r/E,
\end{aligned}
\tag{2}
$$

where $\mathbf{n} = (n_1, n_2)$ is the outward normal to S, E is the equivalence relation

$$
\mathbf{g} \sim \mathbf{g}', \qquad \text{iff} \quad \mathbf{g}(\mathbf{x}) - \mathbf{g}'(\mathbf{x}) = c\ \mathbf{n}(\mathbf{x}), \quad \text{for} \quad c \in R.
$$

In order to derive an equivalent variational formulation for problem (1), we need an integral representation for the solution for problem (1). Let $(\mathbf{v}, q)$ and $(\mathbf{v}^*, q^*)$ be any functions. The Green formula for problem (1) over a domain G is as follows:

$$
\begin{aligned}
2\nu \sum_{i,j=1}^{2} \int_G e_{ij}(\mathbf{v}) e_{ij}(\mathbf{v}^*)\ d\mathbf{y} + \int_G \mathbf{v}^* \cdot \left(\nu \Delta \mathbf{v} - \nabla q\right)\ d\mathbf{y} \\
- \int_G q \nabla \cdot \mathbf{v}^*\ d\mathbf{y} = \int_{\partial G} \mathbf{n} \cdot \left(\mathcal{T}(\mathbf{v}, q)\mathbf{v}^*\right)\ ds(\mathbf{y}),
\end{aligned}
\tag{3}
$$

where $\mathcal{T}(\mathbf{u}, p) = -p\mathcal{I} + 2\nu\mathcal{E}(\mathbf{u})$ is the stress, $\mathcal{I}$ is the identity, and $\mathcal{E}(\mathbf{u}) = \left[e_{ij}(\mathbf{u})\right]_{i,j=1}^{2}$ is the deformation-rate with $e_{ij}(\mathbf{u}) = \frac{1}{2}\left(\frac{\partial u_i}{\partial x_j} + \frac{\partial u_j}{\partial x_i}\right)$.

Switching $(\mathbf{v}, q)$ and $(\mathbf{v}^*, q^*)$ in equation (3), we have

$$\begin{aligned} &\int_G \left[\mathbf{v}\left(\nu\Delta\mathbf{v}^* - \nabla q^*\right) - \mathbf{v}^*\left(\nu\Delta\mathbf{v} - \nabla q\right)\right] d\mathbf{y} \\ &\qquad = \int_{\partial G} \left[\mathbf{n}\cdot\left(\mathcal{T}(\mathbf{v}^*, q^*)\mathbf{v}\right) - \mathbf{n}\cdot\left(\mathcal{T}(\mathbf{v}, q)\mathbf{v}^*\right)\right] ds(\mathbf{y}). \end{aligned} \tag{4}$$

The fundamental solution $(\mathcal{U}, \mathbf{P})$ for Stokes problem, where $\mathcal{U} = (\mathbf{U}_1, \mathbf{U}_2) = \left[U_{ij}\right]_{i,j=1}^{2}$ and $\mathbf{P} = (P_1, P_2)$, is given by Ladyzhenskaya [4]:

$$\begin{cases} U_{ij}(\mathbf{x}) &= \dfrac{1}{4\pi\nu}\left[\delta_{ij}\,\ln\dfrac{1}{|\mathbf{x}|} + \dfrac{x_i\, x_j}{|\mathbf{x}|^2}\right], \\ P_i(\mathbf{x}) &= \dfrac{x_i}{2\pi|\mathbf{x}|^2}. \end{cases} \tag{5}$$

Using equation (4) and the fundamental solution (5), we can derive an integral representation for the solution $(\mathbf{u}, p)$ of problem (1), where $(\mathbf{u}, p) \in \mathbf{W}(R^2) \times L^2(R^2)$, in the following way. For any $\mathbf{x} \in R^2$, we take a ball $B_\lambda = B(\mathbf{x}, \lambda)$, with the radius λ and the center $\mathbf{x}$, and another ball $B_\Lambda = B(0, \Lambda)$, with the radius Λ and the center origin such that $B_\Lambda \supset (\Omega \cup B_\lambda)$. In equation (4), we choose $(\mathbf{v}, q) = (\mathbf{u}, p)$ and $(\mathbf{v}^*, q^*) = (\mathbf{U}_i(\mathbf{x}-\mathbf{y}), \mathbf{P}_i(\mathbf{x}-\mathbf{y}))$ $(i = 1, 2)$, and then apply equation (4) over Ω and $\Omega' \cap B_\Lambda$. Letting $\lambda \to 0$ and $\Lambda \to \infty$, we obtain the representation, in terms of a simple layer potential, as follows:

$$\begin{cases} \mathbf{u}(\mathbf{x}) &= \displaystyle\int_S \mathcal{U}(\mathbf{x}-\mathbf{y})\cdot\mathbf{g}(\mathbf{y}; \mathbf{u}, p)\, ds(\mathbf{y}) + \mathbf{c}, \quad \mathbf{x} \in R^2, \\ p(\mathbf{x}) &= \displaystyle\int_S \mathbf{P}(\mathbf{x}-\mathbf{y})\cdot\mathbf{g}(\mathbf{y}; \mathbf{u}, p)\, ds(\mathbf{y}), \qquad \mathbf{x} \in R^2/S, \end{cases} \tag{6}$$

where $\mathbf{g}(\mathbf{y}; \mathbf{u}, p) = (\mathcal{T}(\mathbf{u}(\mathbf{y}), p(\mathbf{y}))\,\mathbf{n})\,|_{\text{interior}} - (\mathcal{T}(\mathbf{u}(\mathbf{y}), p(\mathbf{y}))\,\mathbf{n})\,|_{\text{exterior}}$ represents the jump of $\mathcal{T}(\mathbf{u}, p)\mathbf{n}$, the normal stress, across S. $\mathbf{c} \in R^2$ is a constant vector.

We are now in a position to derive an equivalent variational formulation for problem (1). We choose $(\mathbf{v}, q) = (\mathbf{u}, p)$ in equation (3), write out the corresponding equation (3) for G to be Ω and Ω', and then sum up these two formulae together. We obtain

$$2\nu \sum_{i,j=1}^{2} \int_{R^2} e_{ij}(\mathbf{u}) e_{ij}(\mathbf{v}^*)\, d\mathbf{y} = \int_S \mathbf{g}(\mathbf{y};\mathbf{u},p)\cdot \mathbf{v}^*\, ds(\mathbf{y}), \qquad \forall \mathbf{v}^* \in \mathbf{W}(R^2). \tag{7}$$

Noting that $\nabla \cdot \mathbf{u} = \nabla \cdot \mathbf{v}^* = 0$, we can rewrite equation (7) in the form

$$\nu \int_{R^2} \nabla \mathbf{u} \cdot \nabla \mathbf{v}^*\, d\mathbf{y} = \int_S \mathbf{g}(\mathbf{y};\mathbf{u},p)\cdot \mathbf{v}^*\, ds(\mathbf{y}), \qquad \forall \mathbf{v}^* \in \mathbf{W}(R^2). \tag{8}$$

Choosing $\mathbf{v}^* = \mathbf{c} \in R^2$ in equation (8) yields

$$\int_S \mathbf{g}(\mathbf{y};\mathbf{u},p)\cdot \mathbf{c}\, ds(\mathbf{y}) = 0, \qquad \forall \mathbf{c} = (c_1, c_2) \in R^2. \tag{9}$$

Thus, if $(\mathbf{u}, p)$ is the solution of problem (1), then $g(\mathbf{x};\mathbf{u},p)$ must satisfy the constraint (9). Moreover, since the pressure p can only be uniquely determined up to a constant in Ω, from the expressions of $\mathcal{T}(\mathbf{u},p)$ and $\mathbf{g}(\mathbf{x};\mathbf{u},p)$, $\mathbf{g}(\mathbf{x};\mathbf{u},p)$ can only be determined up to a vector $c\,\mathbf{n}$. With these observations and the representation (6), it is easy to see that the following integral equation

$$\mathbf{u}_0(\mathbf{x}) = \int_S \mathcal{U}(\mathbf{x}-\mathbf{y})\cdot \mathbf{g}(\mathbf{y};\mathbf{u},p)\, ds(\mathbf{y}) + \mathbf{c}, \quad \mathbf{x} \in R^2, \tag{10}$$

defines an isomorphism from $\mathbf{H}_0^{1/2}/R^2$ to $\bar{\mathbf{H}}_E^{-1/2}$.

Using equation (10), we can state the equivalent variational formulation as follows: find $\mathbf{g} \in \bar{\mathbf{H}}_E^{-1/2}$ such that

$$\mathcal{A}(\mathbf{g},\mathbf{g}^*) = \mathcal{F}(\mathbf{g}^*), \qquad \forall \mathbf{g}^* \in \bar{\mathbf{H}}_E^{-1/2}, \tag{11}$$

where

$$\begin{aligned} \mathcal{A}(\mathbf{g},\mathbf{g}^*) &= \int_S \int_S \mathbf{g}^*(\mathbf{y}) \cdot \big(\mathcal{U}(\mathbf{x}-\mathbf{y})\mathbf{g}(\mathbf{x})\big)\, ds(\mathbf{y}) ds(\mathbf{x}), \\ \mathcal{F}(\mathbf{g}^*) &= \int_S \mathbf{u}_0 \cdot \mathbf{g}^*\, ds(\mathbf{x}). \end{aligned} \tag{12}$$

THEOREM 1. The $\mathcal{A}(\cdot,\cdot)$ defined in equation (12) is a symmetric, bounded and coercive bilinear form on $\bar{\mathbf{H}}_E^{-1/2} \times \bar{\mathbf{H}}_E^{-1/2}$. The $\mathcal{F}(\cdot)$ defined in equation (12) is a bounded linear functional on $\bar{\mathbf{H}}_E^{-1/2}$. That is, there exist two positive constants C_1, C_2, such that for all $\mathbf{g}$, $\mathbf{g}^* \in \bar{\mathbf{H}}_E^{-1/2}$,

$$\begin{aligned} \mathcal{A}(\mathbf{g},\mathbf{g}^*) &= \mathcal{A}(\mathbf{g}^*,\mathbf{g}), \\ \mathcal{A}(\mathbf{g},\mathbf{g}) &\geq C_1 \left\|\mathbf{g}\right\|^2_{\mathbf{H}_E^{-1/2}}, \\ \left|\mathcal{A}(\mathbf{g},\mathbf{g}^*)\right| &\leq C_2 \left\|\mathbf{g}\right\|_{\mathbf{H}_E^{-1/2}} \left\|\mathbf{g}^*\right\|_{\mathbf{H}_E^{-1/2}}, \\ \left|\mathcal{F}(\mathbf{g}^*)\right| &\leq C_2 \left\|\mathbf{g}^*\right\|_{\mathbf{H}_E^{-1/2}}. \end{aligned} \tag{13}$$

For $\mathbf{u}_0 \in \mathbf{H}_0^{1/2}$, the variational formulation (11) has a unique solution $\mathbf{g} \in \bar{\mathbf{H}}_E^{-1/2}$ that continuously depends on the given data on the right-hand side of (11).

Proof: For any $\mathbf{g}$, $\mathbf{g}^* \in \bar{\mathbf{H}}_E^{-1/2}$, let $\mathbf{u}$ and $\mathbf{u}^*$ be the solutions of variational problem (8) corresponding to $\mathbf{g}$ and $\mathbf{g}^*$, respectively. Thus, both $\mathbf{u}$ and $\mathbf{u}^*$ satisfy equation (10). Thus, equations (10) and (12) yield

$$\begin{aligned} \mathcal{A}(\mathbf{g},\mathbf{g}^*) &= \int_S \int_S \mathbf{g}^*(\mathbf{y}) \cdot \left(\mathcal{U}(\mathbf{x}-\mathbf{y})\mathbf{g}(\mathbf{x})\right) \, ds(\mathbf{y}) ds(\mathbf{x}) \\ &= \int_S \mathbf{g}^*(\mathbf{x}) \cdot \mathbf{u}(\mathbf{x}) \, ds(\mathbf{x}) = \int_S \mathbf{g}(\mathbf{y}) \cdot \mathbf{u}^*(\mathbf{y}) \, ds(\mathbf{y}) \\ &= \nu \int_{R^2} \nabla \mathbf{u}(\mathbf{x}) \cdot \nabla \mathbf{u}^*(\mathbf{x}) \, d\mathbf{x}. \end{aligned} \tag{14}$$

Formula (14) shows that $\mathcal{A}$ is symmetric and bounded. Moreover, we have

$$\mathcal{A}(\mathbf{g},\mathbf{g}) = \nu \int_{R^2} \left|\nabla \mathbf{u}(\mathbf{x})\right|^2 \, d\mathbf{x} \geq C \left\|\mathbf{u}\right\|^2_{\left(W_0^1(R^2)\right)^2/R^2} \geq C_1 \left\|\mathbf{g}\right\|^2_{\bar{\mathbf{H}}_E^{-1/2}}. \tag{15}$$

It is easy to see that the right-hand side of variational formulation (11) is a bounded linear functional on $\bar{\mathbf{H}}_E^{-1/2}$. Thus, by Lax-Milgram theorem we finish the proof.

Variational problem (11) requires the finite-element space to satisfy the constraint (9). It is inconvenient computationally. In order to get rid of the constraint, we introduce a modified variational formulation in form of a saddle-point problem: find $(\mathbf{g}, \mathbf{c}) \in \mathbf{H}_E^{-1/2} \times R^2$ such that

$$\begin{aligned} \mathcal{A}(\mathbf{g}, \mathbf{g}^*) + \mathcal{B}(\mathbf{g}^*, \mathbf{c}) &= \mathcal{F}(\mathbf{g}^*), \qquad & \forall \mathbf{g}^* &\in \mathbf{H}_E^{-1/2}, \\ \mathcal{B}(\mathbf{g}, \mathbf{c}^*) &= 0, \qquad & \forall c^* &\in R^2, \end{aligned} \tag{16}$$

where $\mathcal{B}(\mathbf{g}, \mathbf{c})$ is given below

$$\mathcal{B}(\mathbf{g}, \mathbf{c}) = \int_S \mathbf{g} \cdot \mathbf{c} \; ds. \tag{17}$$

THEOREM 2. For $\mathbf{u}_0 \in \mathbf{H}_0^{1/2}$, the variational problem (16) has a unique solution $(\mathbf{g}, \mathbf{c}) \in \mathbf{H}_E^{-1/2} \times R^2$ that continuously depends on the given data on the right-hand side of the variational problem (16).

Proof: First we verify the well-known inf-sup condition, see Girault [3]. That is, there is a constant $C_3 > 0$ such that

$$\inf_{\mathbf{g} \in \mathbf{H}_E^{-1/2}} \sup_{\mathbf{c} \in R^2} \frac{\mathcal{B}(\mathbf{g}, \mathbf{c})}{\left\|\mathbf{g}\right\|_{\mathbf{H}_E^{-1/2}} \left|\mathbf{c}\right|_{R^2}} \geq C_3 > 0. \tag{18}$$

For any $\mathbf{c} \in R^2$, if we choose $\mathbf{g}^{\mathbf{c}} = \mathbf{c}$, we have

$$\mathcal{B}(\mathbf{g}^{\mathbf{c}}, \mathbf{c}) = \left|\mathbf{c}\right|^2 \text{Meas } (S). \tag{19}$$

Since any norms on R^2 are equivalent, we have

$$\left|\mathbf{c}\right|^2 \text{Meas } (S) = \left\|\mathbf{g}^{\mathbf{c}}\right\|^2_{(L^2(S))^2} \geq C \left\|\mathbf{g}^{\mathbf{c}}\right\|^2_{\mathbf{H}^{-1/2}}. \tag{20}$$

Thus, we have proven the inf-sup condition (18). Combining the second inequality in (13) and the inf-sup condition (18), we finish the proof of the theorem.

BOUNDARY ELEMENT APPROXIMATION

We assume the boundary surface S can be represented as $S = \cup_{i=1}^N S_i$, where $S_i = \mathcal{F}_i\big([0,1]\big)$, $\mathcal{F}_i$ is a smooth bijection that maps the unit interval $[0,1]$

onto S_i. We partition $[0,1]$ into a finite number N_i of elements G_{ij} with length h_{ij}. We assume the partition is quasi-uniform, i.e., $h/\min_i h_i \le C < \infty$, where $h = \max_i h_i$. We construct finite-element spaces $\tilde{\mathbf{V}}^h$ on S as follows:

$$\textbf{Case 1}: \quad \tilde{\mathbf{V}}^h = \Big\{\mathbf{g}^h(\mathbf{x});\ \ \mathbf{g}^h(\mathbf{x}) \in \big(L^2(S)\big)^2,\ \mathbf{g}^h(\mathbf{x})\big|_{\mathcal{F}_i(G_{ij})} \in \big(P_0\big)^2, \quad j = 1,\dots,N_i;\ i = 1,\dots,N.\Big\},$$

$$\textbf{Case 2}: \quad \tilde{\mathbf{V}}^h = \Big\{\mathbf{g}^h(\mathbf{x});\ \ \mathbf{g}^h(\mathbf{x}) \in \big(C(S)\big)^2,\ \mathbf{g}^h\big(\mathcal{F}_i(\mathbf{z})\big)\big|_{G_{ij}} \in \big(P_m\big)^2, \quad j = 1,\dots,N_i;\ i = 1,\dots,N.\Big\}, \tag{21}$$

where P_m is the mth degree polynomial space. We also define the following spaces:

$$\begin{aligned} \mathbf{V}^h &= \tilde{\mathbf{V}}^h \oplus \{\mathbf{n}\}, \qquad \mathbf{V}^h_E = \mathbf{V}^h/E, \\ \bar{\mathbf{V}}^h_E &= \Big\{\mathbf{g}^h(\mathbf{x}) \in \mathbf{V}^h_E;\ \int_S \mathbf{g}^h \cdot \mathbf{c}\, ds = 0, \qquad \forall \mathbf{c} = (c_1, c_2) \in R^2\Big\}. \end{aligned} \tag{22}$$

Having defined the finite-element spaces, we can present a BEM scheme as follows: find $(\mathbf{g}^h, \mathbf{c}) \in \mathbf{V}^h_E \times R^2$ such that

$$\begin{aligned} \mathcal{A}(\mathbf{g}^h, \mathbf{g}^{h*}) + \mathcal{B}(\mathbf{g}^{h*}, \mathbf{c}) &= \mathcal{F}(\mathbf{g}^{h*}), \qquad \forall \mathbf{g}^{h*} \in \mathbf{V}^h_E, \\ \mathcal{B}(\mathbf{g}^h, \mathbf{c}^*) &= 0, \qquad \forall \mathbf{c}^* \in R^2. \end{aligned} \tag{23}$$

In this scheme, the finite-element space is not subject to the constraint (9). By Theorem 2, the variational problem (23) has a unique solution $(\mathbf{g}^h, \mathbf{c}) \in \mathbf{V}^h_E \times R^2$. This means that we can only determine the numerical solution $\mathbf{g}^h$ up to $c\,\mathbf{n}$. In order to determine $\mathbf{g}^h$ completely, we consider the following two cases:

(1) If $\mathbf{n} \in \tilde{\mathbf{V}}^h$, then $\mathbf{V}^h = \tilde{\mathbf{V}}^h$. We impose an extra condition

$$\int_S \mathbf{g}(\mathbf{x}) \cdot \mathbf{n}(\mathbf{x})\, ds(\mathbf{x}) = 0. \tag{24}$$

That is, we find out $(\mathbf{g}^h, \mathbf{c}) \in \tilde{\mathbf{V}}^h \times R^2$ such that the variational problem (23) and equation (24) hold. We can solve them by the least square method.

(2) If $\mathbf{n} \bar{\in} \tilde{\mathbf{V}}^h$, it is easy to see that the variational problem (23) reduces to the following problem: find $\tilde{\mathbf{g}}^h \in \tilde{\mathbf{V}}^h$ such that

$$\begin{aligned} \mathcal{A}(\tilde{\mathbf{g}}^h, \mathbf{g}^{h*}) + \mathcal{B}(\mathbf{g}^{h*}, \mathbf{c}) &= \mathcal{F}(\mathbf{g}^{h*}), \qquad && \forall \mathbf{g}^{h*} \in \tilde{\mathbf{V}}^h, \\ \mathcal{B}(\tilde{\mathbf{g}}^h, \mathbf{c}^*) &= 0, && \forall c^* \in R^2, \end{aligned} \tag{25}$$

From the above discussion, we know that the variational problem (23) only needs to be solved in $\tilde{\mathbf{V}}^h$ by either the least square method (when $\mathbf{n} \in \tilde{\mathbf{V}}^h$) or by standard methods (when $\mathbf{n} \bar{\in} \tilde{\mathbf{V}}^h$). We can also solve problem (23) by the least square method in any case.

Many papers have discussed the issue about the numerical experiments for Stokes problems as well as for other problems. The numerical experiments show that most of the CPU is used to compute the coefficient matrix, while the CPU for the solution of the algebraic system is a minor part. Obviously, the assembly of the coefficient matrix can be performed in parallel to take advantage of modern computer architecture. The accurate calculations of these singular integrals are very important to the accuracy of the numerical solutions. Fortunately, many efficient methods have been developed to solve these problems, *e.g.*, methods of transformation, methods of special solutions. Since they are well-known, we do not discuss these issues any further here.

ERROR ESTIMATES

In this section, we prove an optimal-order error estimate in energy norm and superconvergence estimates in L^∞ norm. From the results by Nedelec [6], we have the following estimates

LEMMA Let R^h be the projection operator from $\left(L^2(S)\right)^2$ onto $\mathbf{V}^h$, then

$$\left\|\mathbf{g} - R^h\mathbf{g}\right\|_{\mathbf{H}^q} \le Ch^{r-q}\left\|\mathbf{g}\right\|_{\mathbf{H}^r}, \tag{26}$$

CASE 1: $-1 \le q \le 0 \le r \le 1$,

CASE 2: $-(m+1) \leq q \leq 1,\ q \leq r,\ 0 \leq r \leq m+1.$

THEOREM 3. Let $(\mathbf{g}, \mathbf{c})$ and $(\mathbf{g}^h, \mathbf{c})$ be the solutions of variational problems (16) and (23) respectively, then we have optimal-order energy norm error estimates for $\mathbf{g} - \mathbf{g}^h$ as follows:

$$\left\|\mathbf{g} - \mathbf{g}^h\right\|_{\mathbf{H}_E^{-r}} \leq C h^{m+1+r} \left\|\mathbf{g}\right\|_{\mathbf{H}^{m+1}}, \quad \frac{1}{2} \leq r \leq m+1. \tag{27}$$

Furthermore, let $(\mathbf{u}, p)$ be given by equation (6), $(\mathbf{u}^h, p^h)$ be defined below:

$$\begin{cases} \mathbf{u}^h(\mathbf{x}) &= \int_S \mathcal{U}(\mathbf{x}-\mathbf{y}) \cdot \mathbf{g}^h(\mathbf{y}; \mathbf{u}, p)\, ds(\mathbf{y}) + \mathbf{c}, \quad \mathbf{x} \in R^2, \\ p^h(\mathbf{x}) &= \int_S \mathbf{P}(\mathbf{x}-\mathbf{y}) \cdot \mathbf{g}^h(\mathbf{y}; \mathbf{u}, p)\, ds(\mathbf{y}), \qquad \mathbf{x} \in R^2/S, \end{cases} \tag{28}$$

then we have an optimal-order energy norm error estimate for $(\mathbf{u}-\mathbf{u}^h, p-p^h)$ as follows:

$$\left\|\left(\mathbf{u}-\mathbf{u}^h, p-p^h\right)\right\|_{(H^1(\Omega))^2 \times L^2(\Omega)/R} + \left\|\left(\mathbf{u}-\mathbf{u}^h, p-p^h\right)\right\|_{(W_0^1(\Omega'))^2 \times L^2(\Omega')}$$
$$\leq C h^{m+3/2} \left\|\mathbf{g}\right\|_{\mathbf{H}_E^{m+1}}. \tag{29}$$

Proof: From equations (16) and (23), we have

$$\mathcal{A}(\mathbf{g}-\mathbf{g}^h, \mathbf{g}^{h*}) = 0, \quad \forall\ \mathbf{g}^{h*} \in \bar{\mathbf{V}}_E^h. \tag{30}$$

$$\begin{aligned} C_1 \left\|\mathbf{g}-\mathbf{g}^h\right\|^2_{\mathbf{H}_E^{-1/2}} &\leq \mathcal{A}(\mathbf{g}-\mathbf{g}^h, \mathbf{g}-\mathbf{g}^h) \\ &= \mathcal{A}\left(\mathbf{g}-\mathbf{g}^h, \mathbf{g}-R^h\mathbf{g}\right) \\ &\leq C_2 \left\|\mathbf{g}-\mathbf{g}^h\right\|_{\mathbf{H}_E^{-1/2}} \left\|\mathbf{g}-R^h\mathbf{g}\right\|_{\mathbf{H}_E^{-1/2}}, \end{aligned} \tag{31}$$

From (26) and (31), we have

$$\left\|\mathbf{g}-\mathbf{g}^h\right\|_{\mathbf{H}_E^{-1/2}} \leq C h^{m+3/2} \left\|\mathbf{g}\right\|_{\mathbf{H}_E^{m+1}}. \tag{32}$$

Thus, the estimate (27) is true for $r = \frac{1}{2}$. As for the case $0 < r < m+1$, we can prove it by the Nitsche technique. The estimate (29) is a direct conclusion of *a priori* estimate for Stokes problem and the estimate (27) with $r = \frac{1}{2}$. Thus, we have finished the proof of Theorem 3.

Theorem 4 gives L^∞ superconvergence estimates for $(\mathbf{u}-\mathbf{u}^h, p-p^h)$ and its derivatives outside a neighborhood of S.

THEOREM 4. Let $(\mathbf{u},p),(\mathbf{u}^h,p^h)$ be defined by eqautions (6) and (28), respectively. For all $\mathbf{x} \in R^2$ with $d = d(\mathbf{x},S) \geq d_1 > 0$, we have the maximum norm estimates:

$$\left|D^\alpha\left(\mathbf{u}-\mathbf{u}^h\right)(\mathbf{x})\right| \leq C(d,\alpha)h^{2m+2}\|\mathbf{g}\|_{\mathbf{H}_E^{m+1}}, \tag{33}$$

$$\left|D^\alpha\left(p-p^h\right)(\mathbf{x})\right| \leq C(d,\alpha)h^{2m+2}\|\mathbf{g}\|_{\mathbf{H}_E^{m+1}}, \tag{34}$$

where $\alpha = (\alpha_1,\alpha_2)$, α_i– nonnegative integers; $d(\mathbf{x},S)$ is the distance between $\mathbf{x}$ and S, $C(d,\alpha)$ is a positive number dependent on $d(\mathbf{x},S)$ and α.

Proof: First, for all $\mathbf{x} \in R^2$ with $d = d(\mathbf{x},S) \geq d_1 > 0$, we have

$$\begin{aligned}\left|D^\alpha\left(\mathbf{u}-\mathbf{u}^h\right)(\mathbf{x})\right| = &\left|\int_S D^\alpha\left(\mathcal{U}(\mathbf{x}-\mathbf{y})\cdot(\mathbf{g}-\mathbf{g}^h)(\mathbf{y})\right)ds(\mathbf{y})\right| \\ &\leq \left\|D^\alpha\mathcal{U}(\mathbf{x}-\mathbf{y})\right\|_{\left(\mathbf{H}^{m+1}\right)^2}\left\|\mathbf{g}-\mathbf{g}^h\right\|_{\mathbf{H}_E^{-(m+1)}}.\end{aligned} \tag{35}$$

Noticing that in this case, we have

$$\left\|D^\alpha\mathcal{U}(\mathbf{x}-\mathbf{y})\right\|_{\left(\mathbf{H}^{m+1}\right)^2} \leq C(d,\alpha), \tag{36}$$

From the estimate (27) with $r = m+1$, (35) and (36), we directly obtain (33). The proof for the estimate (34) can be derived similarly, and so is omitted here.

KEY WORDS: Boundary Elements, Stokes Problem, Convergence Analysis

ACKNOWLEDGEMENTS

This research was supported in part by the funding from the Norwegian Research Council for Science and Humanities, by NSF Grant No. DMS - 8922865, and by the funding from the Institute of Scientific Computations at the University of Wyoming.

References

[1] Brebbia, C. A. *The boundary element method for engineers*, Pentech Press, London, 1978.

[2] Brebbia, C. A., Telles, J. C. F. and Wrobel L. C. *Boundary element techniques*, Springer–Verlag, Berlin and New York, 1984.

[3] Girault, V. and Raviart P.-A. *Finite element methods for Navier–Stokes equations: theory and algorithms*, Lecture Notes in Computational Mathematics **5**, Springer–Verlag, Berlin, 1986.

[4] Ladyzhenskaya, O.A. *The mathematical theory of viscous incompressible flow*, Gorden and Breach, New York, 1969.

[5] Lions, J.L. and Magenes, E. *Non-homogeneous boundary value problems and applications*, Springer-Verlag, Berlin and New York, 1972.

[6] Nedelec, J.C. *Cour de l'ecole d'eté d'analyse numerique*, CEA, IRIA, EPF, 1977.

[7] Temam, R. *Navier-Stokes Equations*, North-Holland, Amsterdam, 1977.

[8] Zhu, J. 'A boundary integral equation method for the stationary Stokes problem in 3D', *Boundary Elements*, (Eds. Brebbia, C.A. *et al.*), pp 283–292, Springer–Verlag, Berlin and New York, 1983.

[9] Wang, H. 'The convergence of BEM for the stationary Stokes problem in three dimensions', *Boundary Elements*, (Ed. Du, Q.), pp. 143–150, Pergamon Press, London, 1986.

The Boundary Element Solution of a Viscous Free Surface Problem

Y. Yuan, D.B. Ingham
Department of Applied Mathematical Studies, The University of Leeds, Leeds LS2 9JT, U.K.

ABSTRACT

In this paper we use the boundary element method to solve a very slow viscous flow with a free surface. An iterative scheme is developed and the numerical solutions are obtained for various non-dimensional fluid flow rates, q, and the non-dimensional surface tension, T.

1. INTRODUCTION

Free surfaces occur in a large range of physical phenomena such as gravity waves, flows through porous media, forces on floating bodies, etc. The free surface problem addressed in this paper is that of a viscous flow moving between two parallel plates which are in relative motion. At an arbitrary point the stationary plate turns through an angle of $\pi/2$ away from the moving plate. At this point the fluid forms a free surface which is extended downstream by the moving plate, see Fig. 1.

The free surface problem is complicated by the fact that the solution domain has an unknown boundary on which conditions are to be imposed. For viscous flows there is, in addition to the kinematic condition, the somewhat more complex requirement of continuity of the stress tensor across the free surface. Numerous authors have successfully tackled such problems using a variety of numerical techniques. Coyne and Elrod [1] using a local perturbation technique dealt with this problem, but the shape of the free surface was restricted to assuming that the contact angle was fixed at $\pi/2$. Further numerical investigations of the problem have been performed by Saito and Scriven [2] and Carter [3]. Both of these works used the finite element technique to solve the problem which arises from slot coating flows and they allowed the contact angle to vary as required by the global solution on the free surface. However, they could not obtain results when the non-dimensional fluid flow rate through the plates was larger than 1/2.

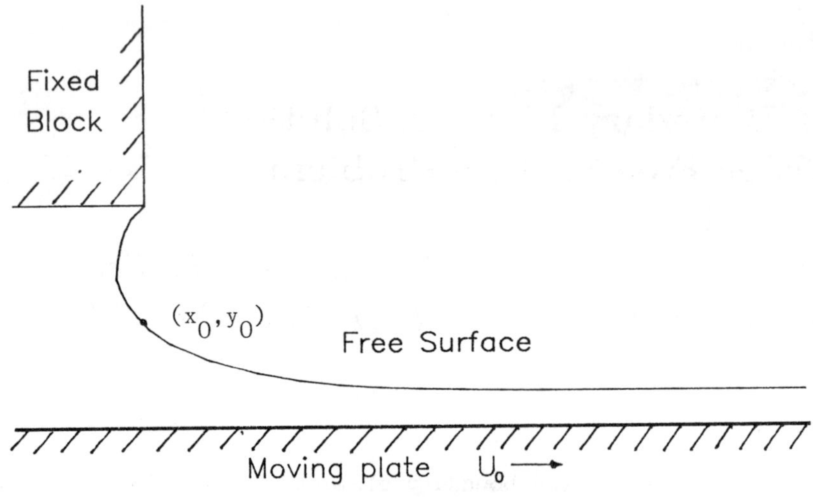

Fig. 1. The geometry of the problem

Liggett [4] and Longuet-Higgins and Cokelet [5] have successfully used the boundary element method (BEM) to obtain solutions to a potential free surface flow problem. In these works the free surface location is obtained by an iterative process based on the satisfaction of the kinematic condition which provides an explicit relationship between known boundary element variables. However, for very slow viscous flows the available conditions provide no such explicit relationship. Ingham and Kelmanson [6] and Ritchie [7] dealt with the viscous flow using the constant BEM in which the viscous flow variables are the stream function, velocity, vorticity and vorticity gradient. Both of these works assumed an analytical function as the fundamental shapes of the free surface and then the problem reduces to the determination of one or more of the coefficients in the series solution. The disadvantage of this method is that it is difficult to find the fundamental shapes of the free surface since they depend on the problem itself. For example, because of the restrictions in the fundamental shapes, Ritchie [7] could not obtain accurate solution when the non-dimensional fluid flow rate was less than 1/2. Ingham and Yuan [8] have presented an iterative scheme in which the free surface was assumed as a collection of points connected by straight line segments. However, they had to restrict the non-dimensional fluid flow rate, q say, such that $1/2 \le q \le 1$. In this paper we have extended the iterative scheme which was introduce by Yuan and Ingham [8] so as to be able to find solutions for $0 < q \le 1$.

In order to obtain more accurate solutions a linear BEM is presented in this paper in which the viscous flows, which are described by the biharmonic equation, are considered. The free surface has been divided into two parts and they are assumed to be functions of x and y, respectively. The reasons for this will be given later in this paper.

2. EQUATIONS AND BOUNDARY CONDITIONS

2.1 GOVERNING EQUATIONS

We consider the two-dimensional incompressible fluid flow between two parallel flat plates, as shown in Fig. 1. The infinite plate moves from left to right at a fixed speed, U_0. At a constant height, h_0, above the moving plate there is a stationary, semi-infinite block. The point on the moving plate directly below the edge of the stationary block is designated as the origin, O. The X-axis lies parallel to the moving plate and the Y-axis is at right angles to the moving plate with Y increasing towards the stationary block. A free surface, Γ, forms at the edge of the stationary block and fluid is swept downstream by the moving plate where the free surface eventually runs parallel to the moving plate at a height, h_d, above the moving plate.

The following non-dimensional quantities are introduced

$$x = X/h_0 \qquad y = Y/h_0 \qquad \mathbf{u} = \mathbf{U}/U_0 \qquad p = Ph_0/\rho\upsilon U_0$$

where υ, ρ and U_0 are the kinematic viscosity, the density and the velocity of the moving plate, respectively. For the steady two-dimensional flow of an incompressible Newtonian fluid and assuming that the Reynolds number, $Re = U_0h_0/\upsilon$, is very small then the Navier-Stokes equation reduces to

$$\nabla p = \nabla^2 \mathbf{u} \tag{2.1}$$

and the continuity equation can be written

$$\nabla \cdot \mathbf{u} = 0 \tag{2.2}$$

On introducing the stream function ψ, the x- and y-components of velocity are then given by

$$u = \psi_y, \qquad v = -\psi_x \tag{2.3}$$

respectively. From equations (2.1-2.3) it may be shown, see Batchelor [10], that ψ satisfies the biharmonic equation

$$\nabla^4 \psi = 0 \tag{2.4}$$

On introducing the vorticity, ω, equation (2.4) may be written in the form

$$\begin{cases} \nabla^2 \psi = \omega & (2.5) \\ \nabla^2 \omega = 0 & (2.6) \end{cases}$$

2.2 BOUNDARY CONDITIONS

The fluid flow far upstream of the origin is constrained to flow between two parallel plates under the action of a pressure gradient and a moving plate, i.e. as $x \longrightarrow -\infty$ the flow is a Couette-Poiseuille flow which means that the fluid velocity profile across the flow is parabolic. This results in the stream function being a cubic function of y and

$$\left.\begin{array}{l}\psi_y \rightarrow 3\ (1 - 2q)y^2 + 2\ (3q - 2)\ y + 1 \\ \omega \rightarrow 6\ (1 - 2q)y + 6q - 4\end{array}\right\} \text{as} \left\{\begin{array}{l} x \rightarrow -\infty \\ 0 \le y \le 1\end{array}\right. \tag{2.7}$$

If the air resistance at the fluid/gas interface is neglected then downstream of the origin the fluid flow becomes plug flow, i.e. the fluid moves as a body with a uniform speed, U_0, and

$$\psi_y \rightarrow 1, \qquad \omega \longrightarrow 0 \qquad \text{as } x \longrightarrow +\infty, \quad 0 \le y \le q \tag{2.8}$$

On the plates we enforce the no slip condition and thus we have

$$\psi = q, \qquad \psi_y = 0 \qquad \text{on } y = 1,\ x \le 0 \tag{2.9}$$

$$\psi = 0, \qquad \psi = -1 \qquad \text{on } y = 0,\ -\infty < x < \infty \tag{2.10}$$

In order to determine the shape of the free surface three boundary conditions are required, and these are

(i) No fluid crosses the interface.

(ii) There is zero shear stress on the free surface.

(iii) The normal stress is balanced by the pressure and the surface tension.

On the free surface we consider a local right-handed coordinate system defined by the vectors **t** and **n**, where **t** denotes the unit tangent vector and **n** is the unit outward normal to the free surface. Then the boundary conditions on the free surface may be expressed in the form

$$\text{(i)}\ \psi = q \tag{2.11}$$

$$\text{(ii)}\ \psi'' - \psi_{tt} = 0 \tag{2.12}$$

$$\text{(iii)}\ -T\,k = -\,p + \sigma'' \tag{2.13}$$

where q (= Q/h_0U_0, where Q is the fluid flow rate) is a non-dimensional fluid flow rate, (′) denotes the normal derivative, T (= T_0/h_0P_0 , where T_0 is the surface tension) is the non-dimensional surface tension, k is the local non-dimensional curvature of the free surface, p is the non-dimensional pressure and σ is the non-dimensional stress tensor, see Ingham and Kelmanson [6] and Yuan and Ingham [9]. We now eliminate the pressure p from the problem and instead of the t-derivatives, which appear in the boundary conditions, it is more convenient to have derivatives with respect to s, the arc length along the free surface. The boundary conditions on free surface may then be written, see Yuan and Ingham [8] and Kuiken [9],

$$\left.\begin{aligned} \psi &= q \\ \omega &= -\,2k\,\psi' \\ -T\,k_s &= \omega' + 2\,\psi'_{ss} \end{aligned}\right\} \text{ on } \Gamma \tag{2.14}$$

In order to solve the problem using the BEM, a closed boundary is required for the solution domain. Thus boundaries are placed upstream and downstream of the origin and on the upstream boundary the stream function is imposed as a cubic function of y and on the downstream boundary the plug flow conditions are imposed. The upstream and downstream boundaries have been placed at a distance, $-d_1$ and d_2 , from the origin respectively, where d_1 and d_2 are positive constants. The full boundary conditions may now be written in the form

$$\begin{aligned} \psi &= q \\ \psi_y &= 0 \end{aligned} \quad \text{on} \left\{ \begin{aligned} &y = 1 \\ &-d_1 \le x \le 0 \end{aligned} \right. \tag{2.15}$$

$$\begin{aligned} \psi &= 0 \\ \psi_y &= -1 \end{aligned} \quad \text{on} \left\{ \begin{aligned} &y = 0 \\ &-d_1 \le x \le d_2 \end{aligned} \right. \tag{2.16}$$

$$\begin{aligned} \psi &= (1 - 2q)y^3 + (3q - 2)y^2 + y \\ \omega &= 6\,(1 - 2q)y + 6q - 4 \end{aligned} \quad \text{on} \left\{ \begin{aligned} &x = -d_1 \\ &0 \le y \le 1 \end{aligned} \right. \tag{2.17}$$

$$\begin{aligned} \psi &= y \\ \omega &= 0 \end{aligned} \quad \text{on} \left\{ \begin{aligned} &x = d_2 \\ &0 \le y \le q \end{aligned} \right. \tag{2.18}$$

$$\psi = q \quad \text{on } \Gamma \tag{2.19}$$

$$\omega = -\,2k\,\psi' \quad \text{on } \Gamma \tag{2.20}$$

$$-T\,k_s = \omega' + 2\,\psi'_{ss} \quad \text{on } \Gamma \tag{2.21}$$

The boundary conditions shown in (2.17) and (2.18) on the upstream and downstream boundaries are idealised in that they are only correctly posed at an infinite distance from the origin. Hence the values of d_1 and d_2 are increased until any further increase in these values produce results which do not change by more than about 1% everywhere in the solution domain. It is found that taking $d_1 \ge 3$ and $d_2 \ge 5$, if the non-dimensional surface tension is less then 50, causes negligible changes in the results presented in this paper. Because the larger the non-dimensional surface tension that is imposed then the further downstream from the origin it is before the plug flow is developed and therefore the value of d_2

depends on the non-dimensional surface tension. In all the results presented in this paper we have taken $T \leq 50$, and then $d_2 = 5$ may be imposed. We note that in order to balance the fluid flow rate upstream and downstream we must have $0 \leq y \leq q$ downstream.

3. NUMERICAL METHOD

For any given combination of the non-dimensional fluid flow rate, q, and the non-dimensional surface tension, T, the problem reduces to finding the location of the free surface which satisfies all three boundary conditions (2.14). Once the location of the free surface is known then the stream function and vorticity may then be calculated everywhere within the fluid.

The fluid flow rate depends upon the pressure gradient, i.e. depends on the difference between the fluid pressure upstream, P_1, and the ambient pressure downstream, P_2. Yuan and Ingham [8] have presented the results for $1/2 \leq q \leq 1$ but in this paper we concentrate on the parameter range $0 < q \leq 1/2$. In this case the contact angle, α_c, which is defined as the angle between the stationary plate and the free surface at the attachment point Q(0,1), may be less than $\pi/2$. This means that the free surface may have two possible values of y for a given value of x when $x < 0$, see Saito and Scriven [2] and Carter [3]. In order to avoid the multivalued function of x, we take x to be a function of y on the free surface, i.e. $x = \Gamma(y)$. However, because the free surface is parallel to the moving plate far downstream then a very small error in the estimate of y will cause a large change in the value of x. Therefore in order to avoid having free surfaces which are either a multivalued function of x or a sensitive function of y we divide the free surface into two parts, namely, $x = \Gamma_1(y)$, which is defined on $[y_0, 1]$, and $y = \Gamma_2(x)$, which is defined on $[x_0, 5]$, and those two parts are matched at the point (x_0, y_0), see Fig. 1.

We parameterize the free surface as a collection of points connected by N_1 straight line segments. The problem is then solved enforcing on the free surface two of the three boundary conditions and the third boundary condition is used to determine how the free surface should be moved in order to satisfy all three boundary conditions. We have taken the boundary condition (2.20) as the iterative condition and we have found that it is the most appropriate approach in order to ensure the iterative process converges most rapidly. The steps required to find the free surface are detailed as follows:

Step 1. Specify a non-dimensional fluid flow rate, q, and a non-dimensional surface tension, T.

Step 2. Specify an initial free surface, $x = \Gamma_1(y)$ and $y = \Gamma_2(x)$.

Step 3. Solve the boundary value problem using the BEM with the two boundary conditions specified on the free surface. Then we obtain all the values of ψ, ψ', ω and ω' on the boundary of the solution domain.

Step 4. Substitute the boundary values which we have obtained in step 3 into the boundary condition (2.12) on the free surface. Then at each mesh point we calculate the residual

$$R_{i,m} = \omega + 2\ k\ \omega'$$

where m denotes the mth iteration and $i = 1, \ldots, N_1$, and the total residual is given by

$$R_m = \sum_{i=1}^{N1} |\ R_{i,m}\ |$$

Step 5. Consider the equation

$$k = (\ \alpha_1 R_{i,m} - \omega\ )/\ 2\ \psi' \tag{3.1}$$

where k is the local curvature on the free surface and $0 < \alpha_1 < 1$ is the relaxation factor which depends on the total residual R_m. From the formula (3.1) we obtain the new curvature at all of the mesh points on the free surface. Then consider the ordinary differential equations

$$\begin{cases} y'' - k\ (\ 1 + y'^2)^{3/2} = 0 & x \in (x_0, 5) \\ y'|_{x=5} = 0 \\ y\ |_{x=5} = q \end{cases} \tag{3.2}$$

and

$$\begin{cases} x'' - k\ (\ 1 + x'^2)^{3/2} = 0 & y \in (y_0, 1) \\ x|_{y=1} = 0 \\ x|_{y=y_0} = x_0 \end{cases} \tag{3.3}$$

which are solved numerically. Further, we introduce another under relaxation factor α_2 in order to avoid large movements of the free surface between successive iterations. Hence the new approximation to the free surface is

$$y_{i,m} = (1 - \alpha_2)\, y_{i,m-1} + \alpha_2\, \bar{y}_{i,m}$$

and

$$x_{i,m} = (1 - \alpha_2)\, x_{i,m-1} + \alpha_2\, \bar{x}_{i,m}$$

where $\bar{y}_{i,m}$ and $\bar{x}_{i,m}$ are the solutions of the ordinary differential equations (3.2) and (3.3), respectively.

Step 6. The iteration is considered to have converged when both the residuals on the free surface and the change in position of the free surface between successive iterations are sufficiently small. That is

$$R < \varepsilon_1 \tag{3.a}$$

and

$$\sum_{i=1}^{N1} | y_{i,m} - y_{i,m-1} | < \varepsilon_2 \tag{3.5}$$

where $\varepsilon_1 > 0$ and $\varepsilon_2 > 0$ are two small preassigned constants. If both of the expressions (3.4) and (3.5) are satisfied then the process is complete and the free surface location is found, otherwise return to step 3.

In step 5, on solving the two ordinary differential equations (3.2) and (3.3) then the point (x_0, y_0) will have two values, namely (x_0^1, y_0^1) and (x_0^2, y_0^2) say. Thus we take the new position of the point (x_0, y_0) as

$$x_0 = (x_0^1 + x_0^2)/2 \qquad\qquad y_0 = (y_0^1 + y_0^2)/2$$

4. NUMERICAL SOLUTION

The iterative scheme presented in section 3 requires the specification of some parameters, namely the non-dimensional fluid flow rate, q, the non-dimensional surface tension, T, the number of the segments, N_1, the control parameters, ε_1 and ε_2, and the relaxation factors, α_1 and α_2. As an example of the results obtained, we take these parameters as follows

$$T = 1$$

$$q = 0.1,\ 0.2,\ 0.25,\ 0.3 \text{ and } 0.4$$

$$\alpha_1 = \begin{cases} 0.1 & R_m \ge 1 \\ 0.2 & 1 > R_m \ge 0.5 \\ 0.3 & 0.5 > R_m \ge 0.1 \\ 0.5 & 0.1 > R \end{cases}$$

$$\alpha_2 = \begin{cases} 0.1 & R_m \geq 1 \\ 0.2 & 1 > R_m \geq 0.1 \\ 0.4 & 0.1 > R_m \end{cases}$$

$$N_1 = 20$$

$$\varepsilon_1 = 0.01, \qquad \varepsilon_2 = 0.0001$$

For the choice of the parameters ε_1, ε_2, α_1, α_2 and N_1 we have considered many factors, including the need to obtain accurate solutions, the convergence of the iterative process, the convergence rate and the CPU time. We believe that the results presented in this paper are graphically accurate. The initial approximation to the free surface is given by

$$y = y_0 - (y_0 - q)\tanh(x - x_0) \qquad x_0 \leq x \leq 5 \tag{4.1}$$

and

$$x = a\,y^2 + b\,y + c \qquad y_0 \leq y \leq 1 \tag{4.2}$$

where $x_0 = 0$, $y_0 = (1 - q)/4 + q$. For the smooth attachment of the functions (4.1) and (4.2) the coefficients in (4.2) are given by

$$a = 16/(1 - q)(3 - q)$$
$$b = - 4(5 + 3q)/(1 - q)(3 - q)$$
$$c = 4(1 + 3q)/(1 - q)(3 - q).$$

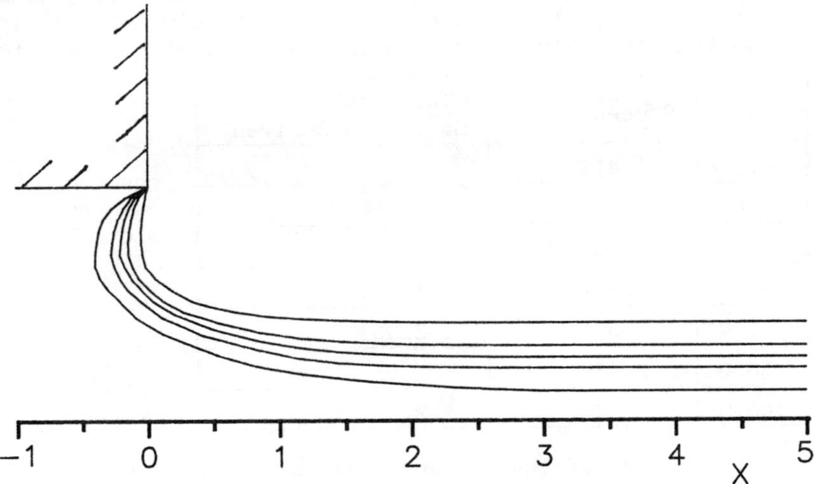

Fig. 2. The free surface profiles as a function of the non-dimensional fluid flow rate, q. Reading from the bottom q = 0.1, 0.2, 0.25, 0.3 and 0.4, respectively.

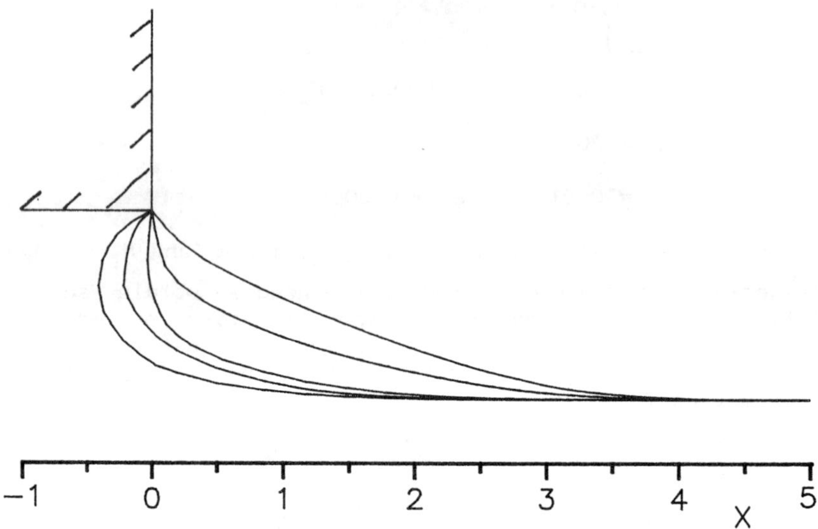

Fig. 3. The free surface profiles as a function of the non-dimensional surface tension, T. Reading from the bottom T = 0.5, 1.0, 2.0, 10 and 50, respectively.

Fig. 2 shows the behavior of the free surface profile as the fluid flow rate varies. These profiles indicate that the contact angle depends upon the fluid flow rate. The free surface profile is flatter at the higher flow rates than at the lower flow rates as one would expect from continuity. We have also investigated the dependence of the free surface profile upon the surface tension, and Fig. 3 shows the solution with q = 0.25 and T = 0.5, 1, 2, 10 and 50, respectively. If T is very large then the solution domain in the x direction has to

T	Present work	The results of Saito and Scriven
0.5	0.415	
1.0	0.702	
2.5	1.294	1.286
5.0	1.772	1.761
10.0	2.076	2.067
50.0	2.428	
100.0		2.641

Table 1. The values of the contact angle as a function of the non-dimensional surface tension, T.

be very large and therefore no results are presented for T > 50 although, in principle, the method should be able to deal with such flows. These profiles show that the local curvature decreases and the contact angle increases as the surface tension increases, again as one would expect from physical consideration. Table 1 shows the values of contact angle for $q = 0.25$ and $T = 0.5$, 1.0, 2.5, 5.0, 10.0 and 50.0, respectively, and the results of Saito and Scriven are also included in Table 1. The agreement of the present work with Saito and Scriven's results is very good.

It is well known that the normal action of the surface tension on the free surface is equal to T k, see Batchelor [10]. Hence if the fluid flow rate is fixed, which means that the difference in the upstream fluid pressure and the downstream ambient pressure, $P_1 - P_2$, is specified, then in order to satisfy the boundary condition (2.13) the local curvature, k, has to decrease as the surface tension increases. On the other hand, if the surface tension is fixed, then the local curvature has to decrease as the fluid flow rate increases. Thus the free surface profile should be flatter at the higher flow rates than at the lower flow rates. The numerical solutions confirm this observation.

In conclusion, in this paper an iterative technique has been developed which is very effective for solving this free surface problem. It is hoped to extend the method described in this paper to deal with the problems in which the fluid flow rate is no longer restricted such that $0 < q \leq 1$. Furthermore, this technique may be used with confidence in other very slow viscous flow problems in which free surfaces occur, e.g the cavitation bubble problem, the sintering problem, etc.

ACKNOWLEDGEMENT

The authors would like to thank the Sino British Friendship Fellowship Scheme for the financial support of Y. Yuan.

REFERENCES

1. Coyne, J.C. and Elrod, H.G. Conditions for the Rupture of a Lubricating Film, Part 1: Theoretical Model, Trans. A.S.M.E. J. of Lub. Tech., Vol. 92, pp 451-456, 1970.

2. Saito, H. and Scriven, L.E. Study of Coating Flows by the Finite Element Method, J. Comp. Phys., Vol. 42, pp 53-76, 1981.

3. Carter, G.C. The Modelling and Analysis of Coating Processing, Ph.D. Thesis, Leeds University, 1985.

4. Liggett, J.A. Location of Free Surface in Porous Media, Trans. A.S.M.E. J. Hyd. Div., Vol. 104, pp 353-365, 1977.

5. Louguet-Higgins, M.S. and Cokelet, E.D. The Deformation of Steep Surface Waves on Water, part 1: A Numerical Method of Computation, Proc. R. Soc. London, Vol. A350, pp 1-26, 1976.

6. Ingham, D.B. and Kelmanson, M.A. Boundary Integral Equation, Analyses of Singular, and Biharmonic Problems, Lecture Notes in Eng., Vol. 7, Springer-Verlag,BerlinHeidbergand New York, 1984.

7. Ritchie, J.A. The Boundary Element Method in Lubrication Analysis, Ph.D. Thesis, Leeds University, 1989.

8. Yuan, Y. and Ingham, D.B., The Numerical Solution of Viscous Flows with a Free Surface, 1st Conf. Comp. Mod. Free and Moving Bound. Prob., Southampton, pp. 325-339, 1991.

9. Kuiken, H.K. Viscous Sintering: The Surface-Tension-Driven Flow of a Liquid from Under the Influence of Curvature Gradients at its Surface, J. Fluid Mechanics, Vol. 14, pp 503-515, 1990.

10. Batchelor, G.K. An Introduction to Fluid Dynamics, Cambridge University Press, 1967.

SECTION 4: SPECIAL FLOW SITUATIONS

3-D Time Dependent Navier-Stokes Solutions with Finite and Boundary Elements

U. Gulcat*

Faculty of Aeronautics and Astronautics, Istanbul Technical University, Maslak 80626, Istanbul, Turkey

ABSTRACT

Full Navier-Stokes equations are solved to study viscous flow past a 3-D body. In time, fractional steps are performed for obtaining the intermediate velocity values from the momentum conservation using the Galerkin Finite Element discretization in space. The Poisson's equation is solved for a derived scalar variable to evaluate the pressure field at each time step. For imposing the far field boundary conditions on pressure approximate boundary element concept external to the finite element region is utilized. Presented are the numerical results obtained for viscous flow past a swept bump on a circular cylinder. The technique proves itself to be capable of depicting the flow field with small number of grid points used in discretization in space and with considerably large steps in time.

INTRODUCTION

In recent years, accessibility to high speed computational tools enabled researchers to solve full Navier-Stokes equations in numerical investigations of general viscous flows of internal or external type. For internal flows the flow domain is finite and it can be discretized with desired number of grid points depending on the accuracy sought for the problem under consideration. However, for the flow past a solid body the fluid domain of concern extends to infinity. Consequently, the prescription of far field boundary conditions with finite computational domain creates problems as regards to the number of grid points. In order to remedy this, the effect of far field is brought on to the finite computational domain with proper type of discrete elements based on a firm theoretical considerations.

In two dimensional external flows a suitable conformal mapping technique can conveniently be applied to map infinity on a finite point in numerical study of

* Presently, Visiting Fulbright Scholar (1991-92) at Purdue University School of Engineering and Technology at Indianapolis, IN 46202-5132

viscous flows; e.g. Mehta[1]. Even more conveniently, without involvement of large values for scale factor of transformation, the far field conditions can be brought onto the finite domain utilizing either infinite elements [2], boundary integrals [3], or in general truncated domain - artificial boundary conditions [4] for solution of two or three dimensional viscous flows. The infinite elements [5] are based on domain integration and are non-isoparametric in nature. The boundary elements are, however, based on boundary integration and if performed directly results in fully populated matrices representing the coupling of all nodes on the boundary. There is a practical approach to avoid this coupling via utilizing the smoothness of the solution at the infinity. This is approximate boundary element approach [6] and is equivalent to the radiation or similar boundary conditions of Poisson's type equations.

In this paper a procedure based on finite and boundary elements is utilized for attaining appreciable reduction in number of grid points in flow domain. In discretization fine mesh close to the body and a coarse mesh away from the body is utilized and at the boundaries reaching to the far field the discrete domain is surrounded with boundary elements. Marching in time the method of fractional step is employed. The Galerkin Finite Element formulation is applied to momentum equation for obtaining the velocity field at fractional time steps. Then the fractional time step velocity values are used in right hand side of the Poisson's equation to evaluate the new time level pressure field from the pressure field of previous time step. With this, the no slip and the far field boundary conditions are imposed properly. In addition, for high Reynolds number flows the option of upwinding on convection terms of momentum equation is possible. The proposed method is applied to study a flow past a circular cylinder with a swept bump. The cylinder is immersed in a uniform stream with its axis being in the flow direction. The flow Reynolds number based on the circular cylinder radius and the free stream speed is 40. After the impulsive start of the cylinder motion the evolution of the flow is studied. The results presented as velocity profiles at various stations on the cylinder indicate that the procedure is capable of depicting the flow field with considerably small number of grid points.

FORMULATIONS

In primitive variables the non dimensional form of the Navier-Stokes Equation reads,

$$\partial \mathbf{u} / \partial t + (\mathbf{u} \cdot \nabla) \mathbf{u} = -\nabla p + (1/R) \nabla^2 \mathbf{u} \tag{1}$$

and the continuity equation

$$\nabla . \mathbf{u} = 0 \tag{2}$$

where $\mathbf{u}$, p, t are nondimensional velocity vector, pressure and time. R is the Reynolds number ∇ is the vector operator. Using the fractional step, similar to that given in Reference [7], with n denoting the time

$$(\mathbf{u}^{n+1} - \mathbf{u}^n) / \Delta t + (\mathbf{u}^n . \nabla) \mathbf{u}^n = - \nabla p^{n+1} + (1/R) \nabla^2 \mathbf{u}^n \tag{3}$$

and

$$(\mathbf{u}^{n+1/2} - \mathbf{u}^{n}) / \Delta t + (\mathbf{u}^{n} . \nabla) \mathbf{u}^{n} = - \nabla p^{n} + (1/R) \nabla^{2} \mathbf{u}^{n} \qquad (4)$$

Subtracting Equation 4 from 3 and introducing new scalar ϕ resuts in

$$\mathbf{u}^{n+1} = \mathbf{u}^{n+1/2} + \nabla \phi \quad \text{with} \quad p^{n+1} = p^{n} - \phi / \Delta t \qquad (5.\ a,b)$$

and because of $\mathbf{u}^{n+1}$ being solenoidal ϕ satisfies

$$\nabla^{2} \phi = - \nabla . \mathbf{u}^{n+1/2} \qquad (6)$$

Here $\mathbf{u}^{n+1/2}$, the fractional step velocity field, is not incompressible.

In Equation 6, when the pressure is specified $\phi = 0$, and when the velocity is specified $\partial\phi/\partial n = 0$ are prescribed as the boundary conditions. With this evaluation of the velocity values become quiet simple at the down stream.

Galerkin Formulation

For explicit solution of fractional step velocity field the Galerkin Finite Element formulation is applied to Equation 4. This results in

$$M_{ij} \mathbf{u}_j^{n+1/2} = M_{ij} \mathbf{u}_j^{n} + \Delta t \,[\, B_{\alpha i} + p_e^{n} C_{\alpha i} - D_{ij} \mathbf{u}_j^{n} - (1/R) S_{ij} \mathbf{u}_j^{n} \,] \qquad (7)$$

where the subscripts i,j denotes the discrete locations α runs like x,y,z coordinates. M_{ij} , S_{ij} and D_{ij} are the mass, stiffness and convection matrices. $C_{\alpha i}$ is the gradient operator and $B_{\alpha i}$ shows the effect of the boundary conditions.

$$S_{ij} \phi_j = E_{xij} u_j^{n+1/2} + E_{yij} v_j^{n+1/2} + E_{zij} w_j^{n+1/2} \qquad (8)$$

with $E_{\alpha ij}$ is the α component of the divergence operator and u,v,w are the velocity components. For numerical evaluation of new time step velocity field the finite element formulation of Equation 5.b is utilized that reads,

$$M_{ij} \mathbf{u}_j^{n+1} = M_{ij} \mathbf{u}_j^{n+1/2} + E_{\alpha ij} \phi_j \ , \ \alpha = x,y,z \qquad (9)$$

The discretization of the domain is made with 8 noded isoparametric brick elements which have tri-linear shape functions. The matrices and the operators in algebraic forms are constructed using these shape functions.

Boundary Elements

The Boundary Elements used here are the three dimensional surface elements which cover the far field boundary of the flow domain discretized with brick elements. In three dimension, the boundary elements depend on two coordinates. Their shape functions are expressed in terms of bi-linear functions to be consistent with the finite element approximation. For numerical evaluation of the integrals one has to express the variables in terms of the local coordinates on the boundary element surface. In Figure 1 shown is the relation between the cartesian system and the ξ, η, ζ local coordinate system on the surface.

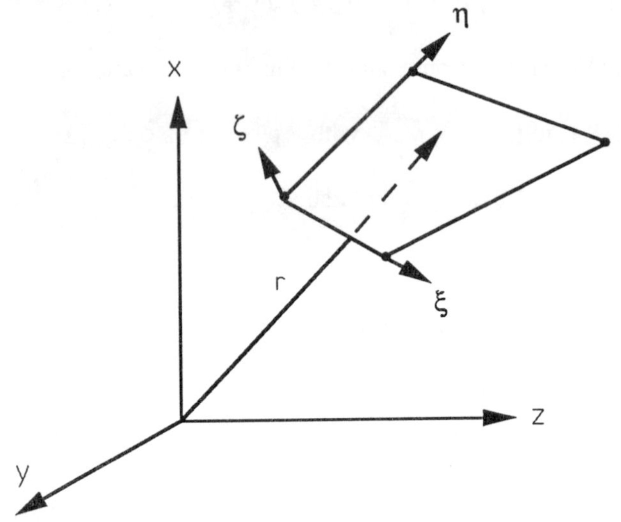

Figure 1. The cartesian and the local coordinates.

The differential area on the surface at location $\mathbf{r} = x\,\mathbf{i} + y\,\mathbf{j} + z\,\mathbf{k}$ in terms of the local coordinates reads,

$$dS = G\, d\xi\, d\eta$$

where G is the reduced Jacobian of the transformation and is the magnitude of the vector normal to the surface. Therefore it reads,

$$G = |\,\mathbf{r}_\xi \times \mathbf{r}_\eta\,| \quad , \quad \mathbf{r}_\xi = x_\xi\,\mathbf{i} + y_\xi\,\mathbf{j} + z_\xi\,\mathbf{k}\,, \quad \mathbf{r}_\eta = x_\eta\,\mathbf{i} + y_\eta\,\mathbf{j} + z_\eta\,\mathbf{k}$$

The explicit expression for G then becomes,

$$G = [\,(y_\xi z_\eta - y_\eta z_\xi)^2 + (x_\eta z_\xi - x_\xi z_\eta)^2 + (x_\xi y_\eta - y_\xi x_\eta)^2\,]^{1/2} \tag{10}$$

After deciding on the boundary element shapes it is straight forward to evaluate the numerical integration utilizing Equation 10.

Approximate Boundary Elements

The coupling of finite and boundary elements result in fully populated matrices because of nature of boundary elements. In order to avoid fully populated matrices one can resort to approximate boundary elements [6] which implies radiation or similar boundary conditions at the far field where the solution behaves in a smooth manner. In three dimensional problems this smoothness at far field can approximately be expressed [6] for unknown ϕ satisfying the Poisson's equation as

$$\partial\phi/\partial r = -\,\phi\,/r \tag{11}$$

If the three dimensional boundary reaching far field is a cylindrical surface as shown in Figure 2 , rather than a spherical the condition (11) then has to be modified as follows. The fundamental solution of Poisson's equation reads

$$\phi^* = 1/(4\,\pi\,r)$$

For the points in the fluid domain near to the cylindrical surface the following integral relation holds

$$\int \phi^* \, \partial\phi/\partial n \, dS = \int \phi \, \partial\phi^*/\partial n \, dS \tag{12}$$

The normal derivative of the fundamental solution considering Figure 2 becomes

$$\partial\phi^*/\partial n = (\partial\phi^*/\partial r)\,(\partial r/\partial n) = (\partial\phi^*/\partial r)\cos\theta \tag{13}$$

Using Equation 13 in (12) one obtains,

$$\int [\partial\phi/\partial n + (\phi/r)\cos\theta]\,dS = 0 \tag{14}$$

For integral equation (14) to be valid on S the integrand vanishes. Therefore,

$$\partial\phi/\partial n = -\,(\phi/r)\cos\theta \tag{15}$$

Equation 14 is the modified form of (11) to be used on cylindrical surfaces.

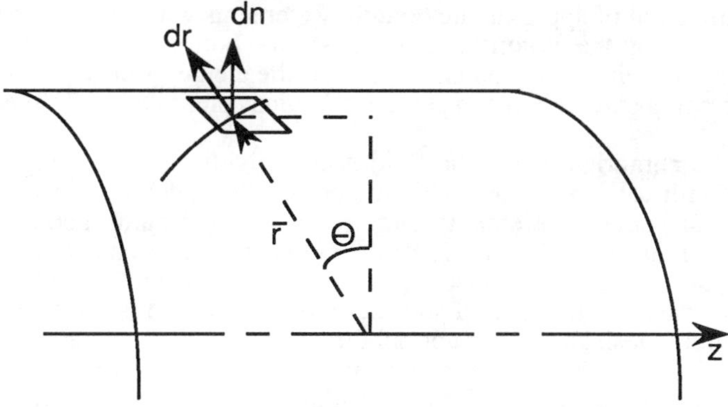

Figure 2. Cylindirical surface S as far field boundary.

Finally, the approximate boundary element approach boils down to satisfying Equation 15 as mixed boundary condition at the finite boundary which represents the far field. This results in altering the form of S matrix in Equation 8 when there is a coupling between the finite and boundary elements at far field boundary [8]. Accordingly, the alteration for the relevant elements of the matrix read,

$$\int N_i N_j \,(\cos\theta)\,/r \, dS$$

where N_i and N_j are the shape functions for the boundary elements.

RESULTS AND DISCUSSION

Flow past a swept-bump on a circular cylinder immersed in a uniform stream aligned with the cylinder axis is studied with numerical procedure based on the above formulation. The sweep on the cylinder simulates a yawed wing. The 10% thick bump geometry is given with its details in Reference [9]. The Reynolds number of 40 is considered for examining the flow with viscous effects dominating the major portion of the flow field.

The fluid motion is impulsively started from the rest , then the flow development in time is solved with the following procedure:

i) From initial velocity and pressure field the fractional step velocities are computed using Equation 7 with finite elements only,
ii) From the fractional step velocity values ϕ is computed solving Equation 8 with coupled finite and boundary elements with far field boundary conditions prescribed with Equation 15,
iii) Using ϕ and the fractional step velocity values new time level velocity and the pressure field is obtained with solution of Equations 9 and 5.

The procedure described above is repeated for the following time steps until the desired time level is reached. The stability criteria for this half explicit scheme can be performed using norm analysis given in Reference [10]. A conservative time increment here is used in order not to violate the stability requirement which is not so stringent.

The utilization of approximate boundary elements with finite elements is first checked solving the potential flow past the bump and the agreement is satisfactory with the results obtained using finite elements only [11] wherein the results are compared with the experimental values presented in Reference [9].

For the numerical study the fluid domain is discretized using 2625 brick elements with total of 3328 nodes including the nodes on the outer surface consisting of boundary elements. Figure 3 shows the bump geometry and the finite element mesh on the surface of the cylinder. The mesh extends eight rows normal to the surface as the spacing gets coarser. The boundary elements are at the last row and for the sake of simplicity only one slice is shown on Figure 3. The minimum mesh spacing in normal direction is 0.05 of the radius close to the surface and the mesh extends to 5 radii away from the surface. There are 16 equally spaced points in the circumferential direction and 26 unequally spaced points in the main flow direction.

On Figure 4 shown is the detailed picture of the bump surface mesh with various stations indicated on it. On those stations, ranging from 1-9 and a-d, the velocity and cross flow velocity profiles will be given. Also indicated on Figure 4 are the free stream velocity and the coordinates for the velocity components. This free stream velocity is applied as the entrance boundary condition at the first station and the far field boundary condition at the last surface composed of nodal points away from the surface of the swept bump . On the last station the velocity profile is computed whereas on u-w plane the symmetry conditions for velocity and pressure are assumed in order not to solve the whole flow domain

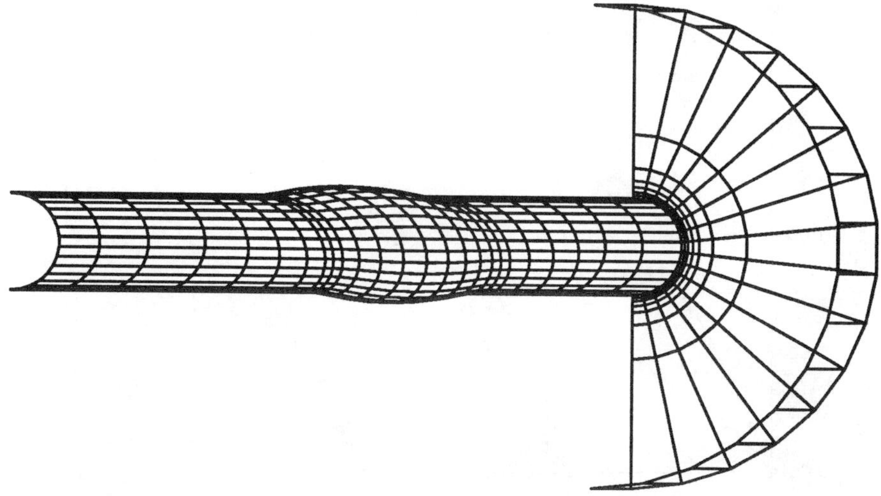

Figure 3. The grid external to the swept bump.

about the cylinder.

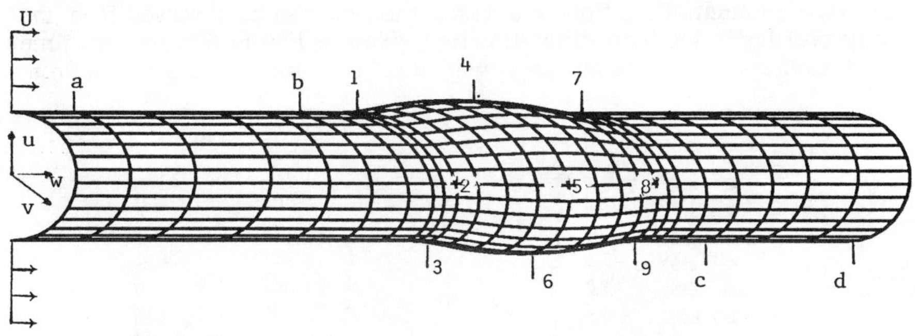

Figure 4. The details of the surface mesh on the bump.

The solution to the impulsively started flow motion is carried out up to the dimensionless time level of $t = 7.5$ which corresponds to the travel of the cylinder almost equal to its length. At about this time level the steady state is almost reached. On Figure 5. shown is the velocity profiles of this time level plotted at various stations on the cylinder surface ranging from a to d and 1, 4 and 7. In this and the following figures the distances from the wall are expressed in terms of the circular cylinder radius $r = 3$. At station a the profile, being next to the free stream velocity profile, still has the highest velocity gradient. At station b and the following station 1, the profiles have tendency to separate. However, at station 4, being at the shoulder of the cylinder, the flow speeds up and avoids the separation. Further downstream flow again slows down. Finally, it reaches a flat region after the bump and regains some speed as can be observed from the comparison of the profiles given at stations c and d.

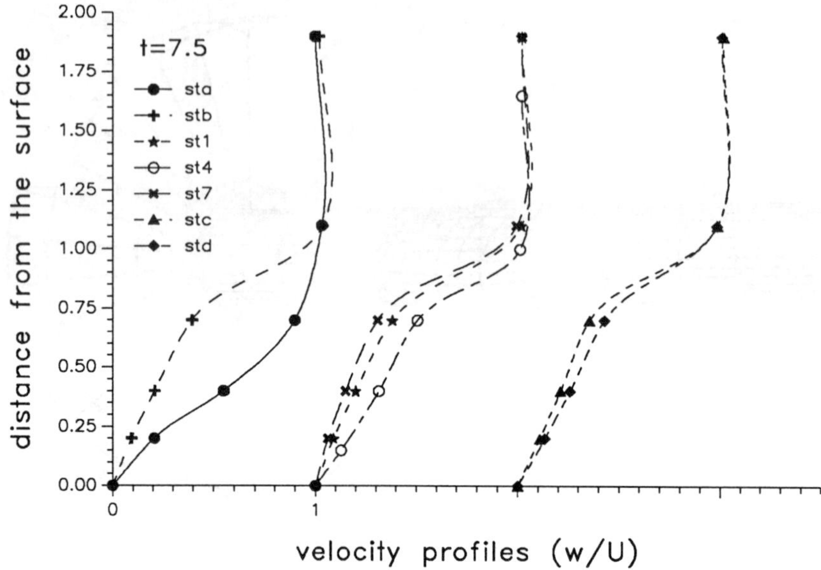

Figure 5. The velocity profiles at time 7.5

The development of the flow with respect to time can be observed from the velocity profiles in the main flow direction given in Figure 6 at various time levels at stations which are on the v-w plane of the surface of the bump. From

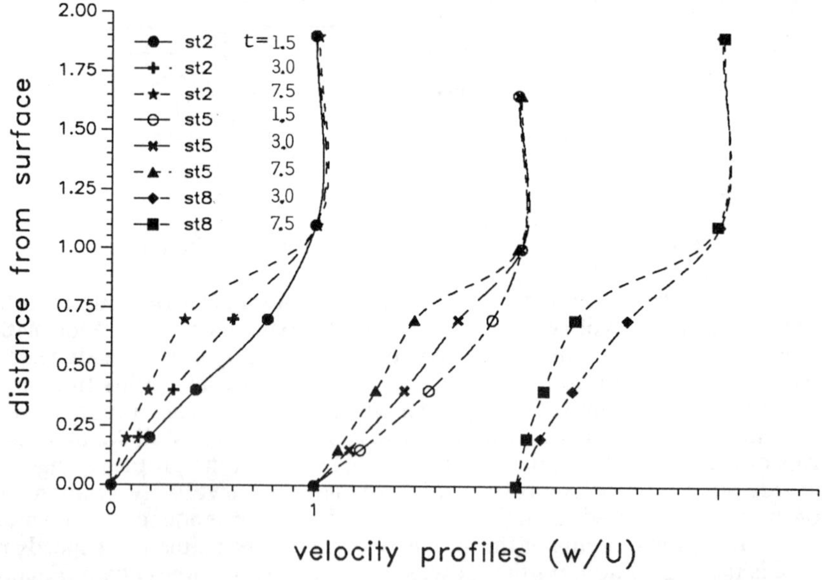

Figure 6. The velocity profiles at various time levels.

Figure 6 at early time level, t = 1.5, the velocity profiles resembles that of two dimensional flow. At later time levels, however, the profiles become less steep and passes inflection points as an indication of the tendency of flow to separate. Since the flow is three dimensional, with the help of the cross flow the flow turns slightly up at the first half of the bump and then turns down towards the bottom of the bump after the shoulder. This is in agreement with the experimental flow visualization given at Reference [9] and can be observed in detail in Figure 7 where the cross flow velocity profile development at the v-w plane for various time levels are shown.

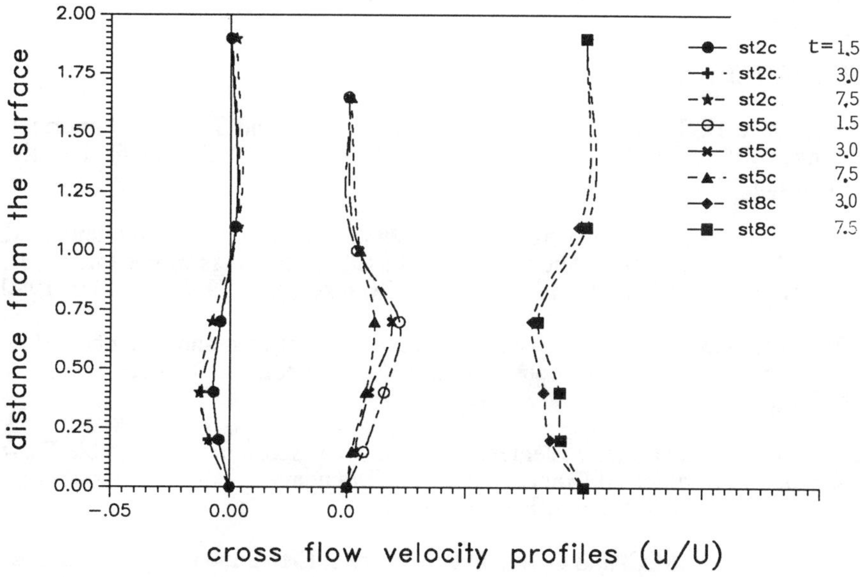

Figure 7. Cross flow velocity at different time levels.

One time step of computation takes about 8 seconds of CPU time on IBM 3090 with vector processing option. The memory requirement is about 6 MByte with single precision arithmetic.

CONCLUSION

A Finite Element Galerkin formulation together with approximate boundary element concept is satisfactorily applied for numerical study of a viscous flow past a swept bump on a circular cylinder. The boundary elements are utilized in presentation of far field boundary condition in order to reduce the number of points used in discretizetion. The results indicate that the procedure is capable of predicting the three dimensional aspects of the flow with appreciable reduction in grid points. The flow features depicted are in qualitatively good agreement with experiment.

The procedure is prepared to solve also the flows with higher Reynolds numbers utilizing the concept of upwinding on convective terms. The next step, hence, is the application of the code for general viscous flows with high

Reynolds numbers where the viscous effects will be confined closer to the solid body surface.

ACKNOWLEDGEMENTS

The full support from Fulbright Scholar Program and The Council for International Exchange of Scholars are greatly acknowledged. Moreover, the computational facilities and the environment of CFD Laboratory of Mechanical Engineering Department at IUPUI made this work possible. Also, the skilled labor from R.U. Payli in producing the hard copies of computer plots is appreciated.

REFERENCES

1. Mehta, U.B. Starting Vortex, Separation Bubbles and Stall - A Numerical Study of Laminar Unsteady Flow Around an Airfoil, Ph.D. Thesis, Illinois Institute of Technology, 1972.

2. Gulcat, U. 3-D Navier-Stokes Solutions with Finite/Infinite Elements (Ed. Pande, G.N. and Middleton, J.) pp 940-946, Proceedings of the Int. Conf. on Numerical Methods in Engineering, Swansea, UK, 1990, Elsevier, 1990

3. Wu, J.C. and Gulcat, U. Separate Treatment of Attached and Detached Flow Regions in General Viscous Flows. AIAA Journal, Vol.19 - 1, pp 20-27, 1981.

4. Gunzberger,M.D. Finite Element Methods for Viscous Incompressible Flows Chapter 16, Truncated Domain - Artificial Boundary Condition Methods, pp 183-190, Academic Press, Boston, 1989.

5. Marquez, J.M.M.C. and Owen, D.R.J. Infinite Elements in Quasi - Static Materially Nonlinear Problems, Computers and Structures, Vol .18, 4, pp 739-751, 1984.

6. Brebbia, C.A. and Dominguez, J. Boundary Elements An Introductory Course, Computational Mechanics, Southampton and Boston, 1989.

7. Mizukami, A. and Tsuchiya, M.A. Finite Element Method for the Three-Dimensional Non-Steady Navier-Stokes Equations, International journal For Numerical Methods in Fluids, Vol.4, pp 349-357, 1984.

8. Aral, M.M. Finite Element Solutions of Selected Partial Differential Equations "Femac Computer Program" METU Faculty of Arts and Sciences, No. 28, 1974.

9. Ozcan, O. Three-component LDA Measurements in a Turbulent Boundary Layer, Exp. in Fluids, Vol.6, pp 327-334, 1988.

10. Temam, R. Navier-Stokes Equation, North-Holland, Amsterdam, 1977.

11. Gulcat, U. and Aslan, R.A. A Finite Element-Finite Difference Solution of a Three Dimensional Viscous-Inviscid Interaction, Commun. appl. numer. methods., Vol.4, 2, pp 251-254, 1988.

Hydrodynamic Properties of Multiple Floating and Submerged Bodies Analysed by a Panel Method

X. Lei, L. Bergdahl

Department of Hydraulics, Chalmers University of Technology, S-412 96 Göteborg, Sweden

Abstract

A computer programme based on a panel method is developed in order to investigate the interaction between waves and a system of floating and submerged bodies. Small-amplitude waves and motions and irrotational flow are assumed and the resulting linear problem is solved in the frequency domain with six degrees of freedom of each body. The developed computer programme is made efficient by using the Haskind relation for the calculation of forces and the symmetry or double symmetry of symmetric configurations.

The computer code is applied to a problem with a wave-energy device, consisting of a floating buoy connected by a pump to a submerged suspended plate. For a single wave-energy device, comparisons are made with results for vertical forces and motions calculated by the method of expansion in matched eigenfunctions. The agreement is good.

As practical applications, the effect of offset between buoy and plate and the effect of the shape of the buoy and plate are investigated.

The programme is coded in FORTRAN 77 and is presently run on a PC 386.

1 BACKGROUND

In the Department of Hydraulics, Chalmers University of Technology, work is conducted concerning the optimum efficiency of a class of wave power devices, consisting of a floating buoy connected to a submerged plate by some type of power take-off mechanism. To that end, a fast semianalytical method was used for the associated analytical problem, by expanding the solution in matched eigenfunctions for the vertical motion of the buoy and plate. Results from this activity were presented earlier by Berggren & Johansson[1] (1992) and Berggren & Bergdahl[2] (1991). A fast method is vital in order to be able to make many calculations, as the applied power take-off mechanism is nonlinear and has to be linearized for each tested sea-state.

Then the question arises how accurate such a solution will be for a real configuration e.g. with buoys of conical shapes instead of buoys with a flat bottom or with offset between the surface buoy and the submerged plate.

To test some of these effects, it was decided to develop Lei's panel programme for multiple floating bodies (Björkenstam & Lei[3], 1990) into a programme also applicable to a configuration of floating and submerged bodies.

In this paper the influence on some of the hydrodynamic properties are presented. The next step would be to evaluate the influence on the power take-off and decide how reliable the fast semianalytical method is for efficiency optimization. This has not yet been done.

2 DEFINITION OF HYDRODYNAMIC PROPERTIES OF THE HOSE PUMP

A buoy riding in waves, connected to a submerged plate by an elastomeric hose, has been proposed as a device for extraction of energy from waves (Hagerman[4], 1988). This device is known as the Hose Pump. The elastomeric hose acts as a pump that is driven by the relative motion between the buoy and the submerged plate. The submerged plate is moored to the sea floor. Following Hagerman the concept is described as follows. During the passage of a wave crest, the buoy heaves up stretching the hose. The helical pattern of steel reinforcing wires in the hose causes it to constrict as it is stretched, thereby reducing its internal volume. This forces sea water out of the hose pump, through a check valve, and into a collecting line to a turbine. After the wave crest has passed the buoy drops down into the succeeding trough, the hose pump returns to its original length, restoring its diameter to its unstretched value. This increase in internal volume draws water into the hose through another check valve, which is open to the sea. (See Fig. 1)

In order to analyse the dynamics of the wave energy device, properties such as hydrodynamic coefficients have to be calculated. The problem is here formulated linearly which means that the coefficients associated with the harmonic motion of one of the bodies are calculated considering the other bodies as fixed. The radiated waves associated with the body in motion will cause a dynamic pressure. In a multi-body system this pressure will not only be experienced by the body in motion but also by the fixed bodies. This leads to an added mass and damping coefficient of the body in motion as well as cross-terms of added mass and damping of the fixed bodies. In the present paper added mass and potential damping including cross-terms are presented for the cases with vertical motions of any of the bodies.

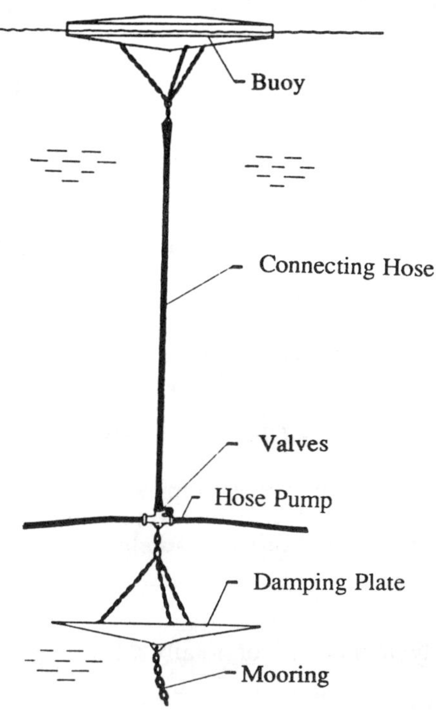

The nomenclature of added mass and potential damping is retained throughout the report also for the cross-terms although this may seem confusing. The added masses of both the fixed bodies and the moving body is then defined as a quantity, that multipled with the acceleration of the moving body gives that part of the hydrodynamic force on the body which is in phase with the acceleration of the moving body. In a similar way the potential damping is defined as a quantity that multipled by the velocity of the moving body gives that part of the hydrodynamic force on the body which is in phase with the velocity of the moving body.

Figure 1. The hose pump concept.
The elastomeric hose acts as a pump, that is driven by the relative heaving motion between a buoy and a submerged plate.

3 SOLUTION TECHNIQUES

In general, there are three major types of solution techniques applicable to the present hydrodynamic boundary value problem;

a) Series expansion in matched eigenfunctions
b) The integral equation method, also known as the panel method
c) The finite element method.

For the purpose of analysing the interaction between multiple floating bodies, Lei[3] (1990) developed a complete programme based on the integral equation approach. This programme has now been developed to take into account also submerged bodies.

Analytical solutions by matched eigenfunction expansions for hydrodynamic interference between the floating buoy and the submerged plate were presented earlier by Berggren & Johansson[1] (1992) and Berggren & Bergdahl[2] (1991). Comparisons with their data were performed and are presented here. Based on the good agreement, a few investigations were performed for more complex and practical problems.

The validation of the presented computer programme was checked in various ways.

a) The accuracy of solution was checked by comparing with the series expansion solution.

b) The numerical convergence was tested by increasing the number of panels.

c) The consistency of the solution was tested by comparing forces calculated by the Haskind relation from the radiation problem with forces calculated by the diffraction theory.

d) The solutions for several devices were checked by setting a large distance between them and comparing with the data for the device.

4 MATHEMATICAL FORMULATION OF THE PROBLEM

The mathematical formulation is based on the assumption that the fluid is incompressible and inviscid and thus the flow irrotational. The waves and the motions of the bodies are harmonic and of small amplitude. The bodies are rigid and of arbitrary shapes.

4.1 The Equation of Motion

The equation of motion of body number m of totally M bodies can be written in the following way.

$$\left\{-\omega^2\left(\left[\mathbf{M}^{m}\right]+\left[\boldsymbol{\mu}^{mm}\right]\right)-i\omega\left[\boldsymbol{\lambda}^{mm}\right]+\left[\mathbf{C}^{m}\right]\right\}\left[\mathbf{X}^{m}\right]$$

$$+\sum_{\substack{j=1\\ j\neq m}}^{M}\left\{-\omega^2\left[\boldsymbol{\mu}^{mj}\right]-i\omega\left[\boldsymbol{\lambda}^{mj}\right]\right\}\left[\mathbf{X}^{j}\right]=\left[\mathbf{F}^{m}\right]$$

$$m, j = 1,2,\ldots,M$$

$\left[\mathbf{M}^{m}\right]$ generalized mass matrix (6×6)

$\left[\mathbf{C}^{m}\right]$ generalized buoyancy restoring force matrix (6×6)

$\left[\mathbf{X}^{m}\right]$, $\left[\mathbf{X}^{j}\right]$ column vectors of generalized amplitudes (6×1) of the generalized displacements of the body

$\left[\boldsymbol{\mu}^{mj}\right]$, $\left[\boldsymbol{\mu}^{mm}\right]$ generalized added mass matrices (6×6) with elements as

$$\mu_{\beta\alpha}^{mj} = \rho \iint_{S_m} \mathrm{Re}\,(\phi_{\beta_j})\, \mathbf{n}_\alpha\, dS$$

$[\lambda^{mj}]$, $[\lambda^{mm}]$ generalized damping matrices (6 × 6) with elements as

$$\lambda_{\beta\alpha}^{mj} = \rho\omega \iint_{S_m} \mathrm{Im}\,(\phi_{\beta_j})\, \mathbf{n}_\alpha \, dS \qquad \alpha = 1,2,...,6 \quad \beta = 1,2,...,6$$

The super- and subscripts denote that the generalized hydrodynamic force of the m:th body's α component results from the j:th motion in the β mode.

$[F^m]$ exciting wave force vector (6 × 1) on body m with components as

$$F_\alpha^m = i\rho\omega \iint_{S_m} (\phi_I + \phi_s)\, \mathbf{n}_\alpha \, dS \qquad \alpha = 1,2,...,6$$

S_m submerged surface of m:th body

4.2 The Potential Problems

The formulation here involves two basic surface wave problems. One is the radiation problem, in which one at a time the bodies are forced to oscillate in stagnant water. The other is the diffraction problem in which all bodies are held fixed in incoming waves. The potential field, ϕ, is governed by the following Laplace equation and boundary conditions. For a definition of the coordinates, see Fig. 2.

$$\nabla^2\phi = 0 \qquad \text{in the fluid}$$

$$\frac{\partial\phi}{\partial z} - \frac{\omega^2}{g}\phi = 0 \qquad \text{in the still water surface } (z = 0)$$

$$\frac{\partial\phi}{\partial z} = 0 \qquad \text{on the bottom } (z = -h)$$

$$\lim_{r \to \infty} \sqrt{r}\left(\frac{\partial\phi}{\partial r} - ik\phi\right) = 0 \qquad \text{outgoing at infinity}$$

where ω is the angular frequency
g the acceleration of gravity
k the wave number

and $i = \sqrt{-1}$

In the radiation problem the fluid velocity perpendicular to the body surface should equal the velocity of the surface which can be formulated as a boundary condition.

$$\frac{\partial\phi}{\partial n} = \mathbf{n} \cdot \delta_{jm}$$

where $\mathbf{n}$ is an outward normal to the surface, δ_{jm} is Kronecker's delta.

In the diffraction problem the normal velocity on the body surfaces is zero.

$$\frac{\partial \phi}{\partial n} = - \frac{\partial \phi_I}{\partial n}$$

where ϕ is the scattered potential and ϕ_I the incident wave potential. The terms on the right hand side of the above two body-surface conditions are known, and they will be at the same position in the respective integral equations. Despite the fact that these two problems are different in physical nature, they are treated by the same numerical procedure.

4.3 Numerical Procedure

Both boundary-value problems are solved by the panel method based on Green's theorem. The potential in the fluid field is controlled by the integration over its closed surface. The Green function here acts as a contributing function, which implies that the Green function G(P,Q) is the potential at the fluid point P(x,y,z) due to a source of unit strength at the surface point Q(ξ,η,ζ).

The Green function given by John[5] (1950) satisfies the free surface condition, the bottom condition, the radiation condition and the Poisson equation ($\nabla^2 G = \delta$). Therefore, the region of source distribution is reduced to body surfaces only. The potential is expressed as

$$\phi(P) = \sum_{m=1}^{M} \iint_{S_m} \sigma(Q) \cdot G(P,Q)\, dS$$

where $\sigma(Q)$ is the source strength. By applying this potential equation on the immersed surfaces and satisfying their kinematic boundary condition, a Fredholm integral equation of the second kind is had as

$$- 1/2\, \sigma(P) + \sum_{m=1}^{M} \iint_{S_m} \sigma(Q) \cdot \frac{\partial G(P,Q)}{\partial n}\, dS = V_n(P)$$

where $V_n(P)$ is the normal velocity ($\mathbf{n} \cdot \delta_{jm}$) in the radiation problem and $- \frac{\partial \phi_I}{\partial n}$ in the diffraction problem.

This integral equation is discretisized by subdividing S_m into an ensemble of panels ΔS. The source strength σ is assumed to be constant on each panel, and the collaboration points are selected as the centroids of the panels. This gives a system of complex linear algebraic equations.

5 THE COMPUTER PROGRAMME

The numerical evaluation of the source strength and the velocity potential in the previous section is achieved by a computer programme written in FORTRAN 77. The discretized formulations are written in the following matrix form

$$\phi = [B] \{\sigma\}$$

$$[E] \{\sigma\} = \{V_n\}$$

where the elements of the matrices [B] and [E] are expressed, respectively, as

$$B(I,J) = \iint_{\Delta S} G(I,J)\, dS$$

$$E(I,J) = \iint_{\Delta S} \frac{\partial G(I,J)}{\partial n}\, dS$$

In the Green function and its normal derivative, there are two singularities, $1/R$ and $\partial(1/R)/\partial n$, when the integral panel is very close to the observed point ($R \rightarrow 0$). In this case, a simple multiplication is not possible. The classical techniques of Faltinsen-Michelsen[6] (1975) and Hess-Smith[7] (1962) are used respectively. Except for these two singularities, the Green function is evaluated by the computer code FINGREEN[8] (1986). The limit for use of the simple approximation is $R > 2\sqrt{\Delta S}$ according to Garrison[9] (1978).

The source strengths $\{\sigma\}$ in the equation system are solved by Gauss elimination. In the present programme code, the Haskind relation and the property of symmetry are included. If the body geometry and the fluid field are symmetrical to one or two vertical planes of symmetry, the size of the problem to be considered will be a half or a quarter of the original one.

6 NUMERICAL CHECKS OF THE PROGRAMME

6.1 Check of consistency by the Haskind relation

The consistency of the developed computer programme for a simple device was tested by using the Haskind relation. That is, the exciting force was calculated both directly from the diffraction potential and indirectly from the radiation potential using the Haskind relation. The geometry of the simple device, consisting of two vertical circular cylinders without offset, are given in Figure 2. The properties are given by

$h_2/h_1 = 0.3$
$R/h_1 = 0.2$
$d_1/h_1 = 0.1$
$d_2/h_1 = 0.1$

As examples of comparisons nondimensional forces in horizontal and vertical directions are shown in Figure 3a for the buoy and in Figure 3b for the plate. The agreement is satisfactory, but for accurate calculations the diffraction solution should be used. The forces are dimensionalized by multiplying with $A\rho g\pi R^2$.

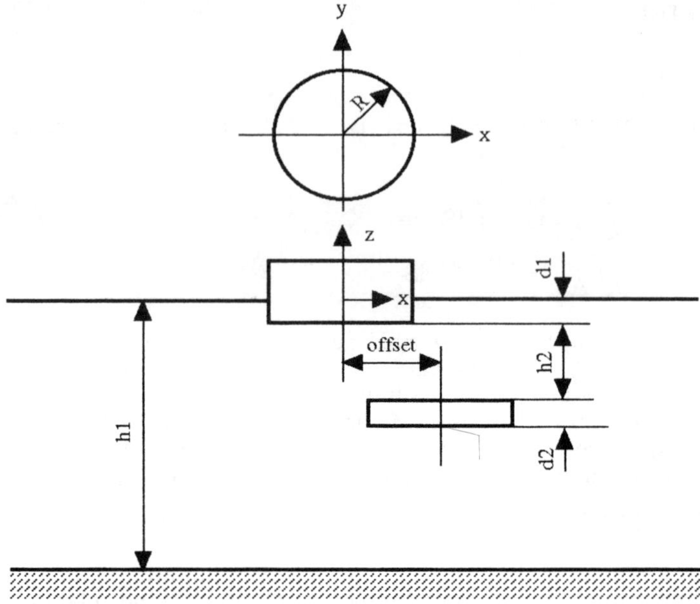

Figure 2 Geometrical properties of the wave energy device.

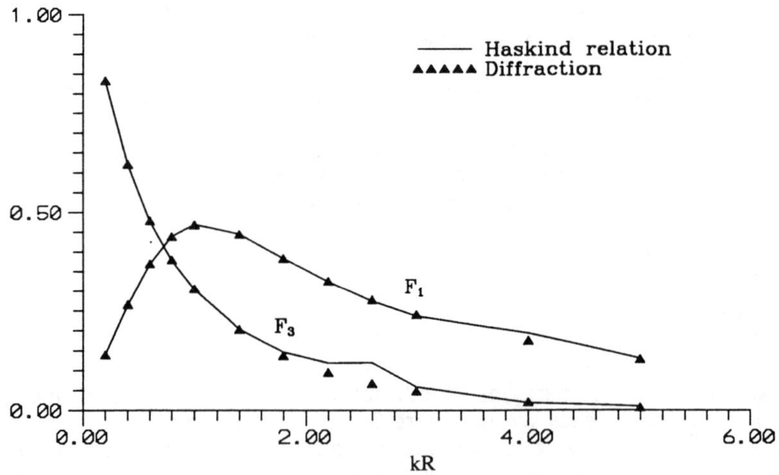

Figure 3a The nondimensional vertical and horizontal exciting forces for the buoy. 450 panels for half of the problem.

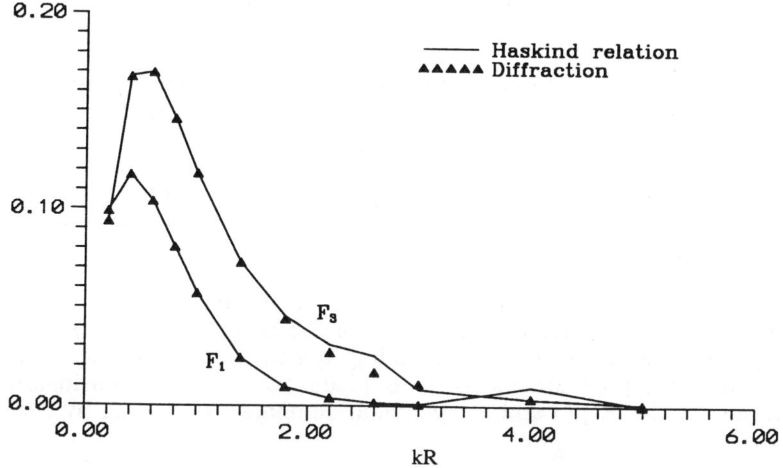

Figure 3b The nondimensional vertical and horizontal exciting forces for the plate. 450 panels for half of the problem.

6.2 Convergence
The convergence for increasing number of panels are shown in Figure 4. The convergence is excellent for the shown diffraction results. For practical purposes 258 panels seem to be enough.

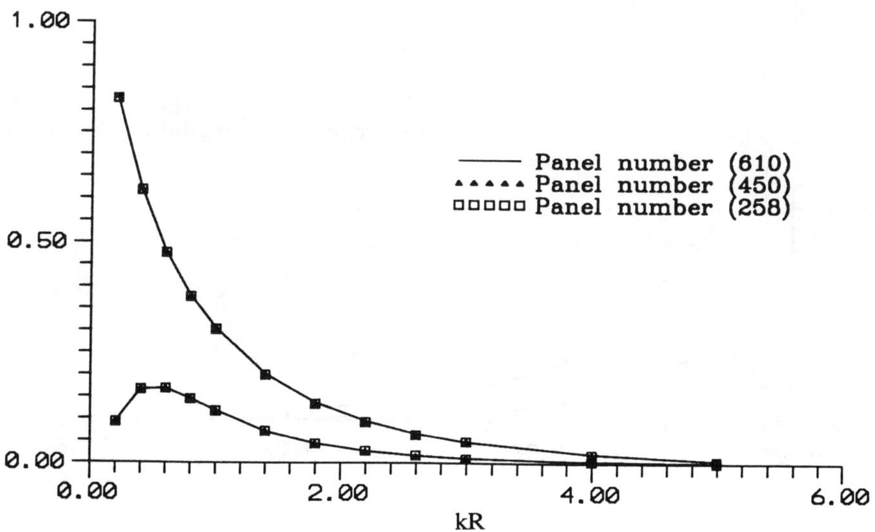

Figure 4 Nondimensional vertical exciting forces for the buoy and submerged plate.

COMPARISON BETWEEN THE TWO METHODS

Some comparisons were made with the results from the method of expansion in matched eigenfunctions by Berggren and Johansson[1] (1992) and Berggren and Bergdahl[2] (1991). The comparisons were made for the same geometry without offset as previously.

In Figure 5 the nondimensional exciting forces for respectively the buoy and the plate are shown for both the present panel method and the eigenfunction method.

In Figures 6a and b the nondimensional added masses and potential damping as a consequence of the motion of the buoy in heave are shown, and in Figures 7a and b the corresponding quantities as a consequence of the motion of the plate are shown. The added mass is nondimensionalized by dividing by $2\,\pi\,R^3\rho/3$ and the potential damping by dividing by $2\,\pi\,R^3\rho\omega/3$.

The agreement between the two methods is satisfactory for all practical purposes both for added mass, potential damping and exciting forces. In Figures 6b and 7a, it can be seen that the coefficients μ_{33}^{12} and η_{33}^{12} are respectively equal to μ_{33}^{21} and η_{33}^{21}. The symmetrical properties of the coefficients are confirmed here.

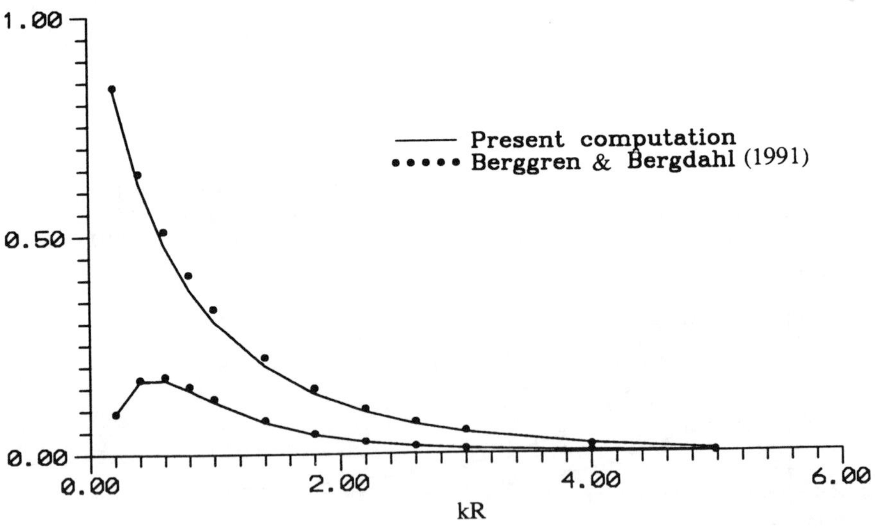

Figure 5 Non-dimensional vertical exciting forces for the buoy and submerged plate. 610 panels for half of the problem.

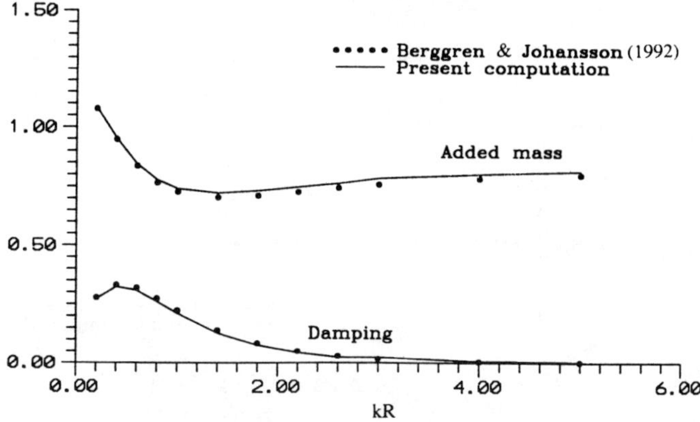

Figure 6a Non-dimensional vertical added mass and damping of the buoy as a consequence of the motion of the buoy

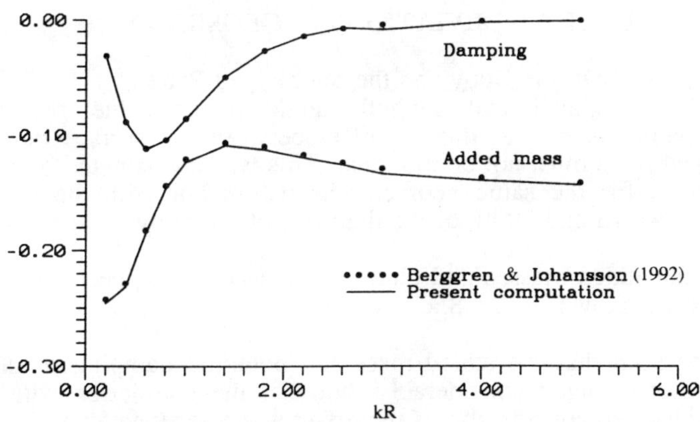

Figure 6b Nondimensional vertical added mass and damping of the plate as a consequence of the motion of the buoy

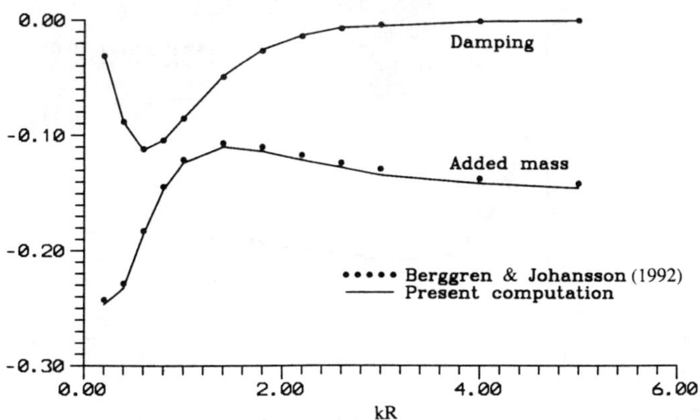

Figure 7a Nondimensional vertical added mass and damping of the buoy as a consequence of the motion of the plate

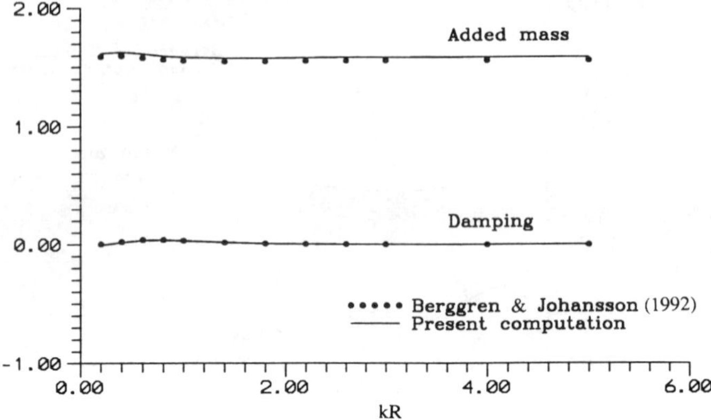

Figure 7b Nondimensional vertical added mass and damping of the plate as a consequence of the motion of the plate

8. INFLUENCE OF DEVIATIONS IN GEOMETRY

8.1 Offset between the Buoy and the Submerged Plate

If the fast semianalytical method should be of value, the change in hydrodynamic properties due to offset between the surface buoy and the submerged plate must not be too great. This was investigated by performing calculations for the same geometry as before but with the vertical axes offset 25, 50, 75 and 100% of the diameter of the buoy.

Results for added masses and damping as a consequence of the motion of the buoy are shown in Fig. 8 a, b, c and d.

It can be seen that the added mass and potential damping of the moving body is not changed considerably, but that the interaction with the other body is changed considerably if the offset is greater than 25%.

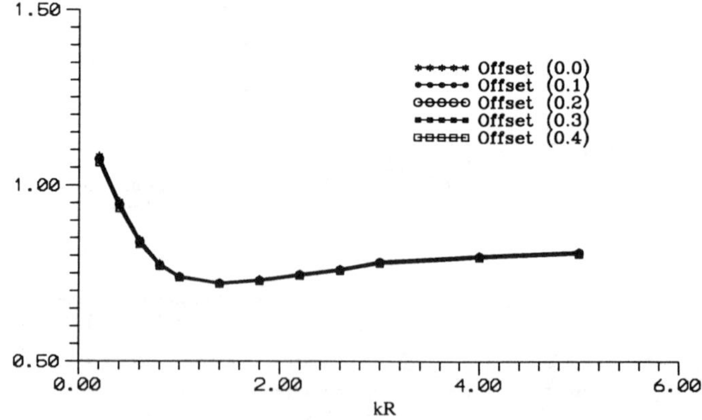

Figure 8a The effect of offset between buoy and plate on the nondimensional vertical added mass for the buoy associated with vertical motion of the buoy

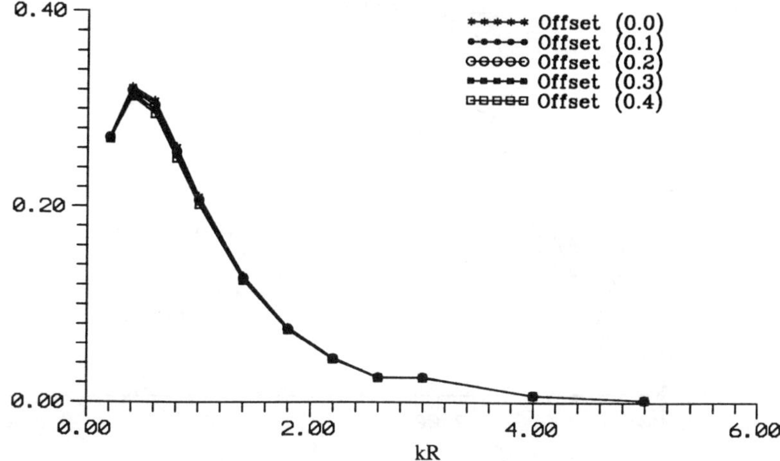

Figure 8b The effect of offset between buoy and plate on the nondimensional vertical potential damping for the buoy associated with vertical motion of the buoy

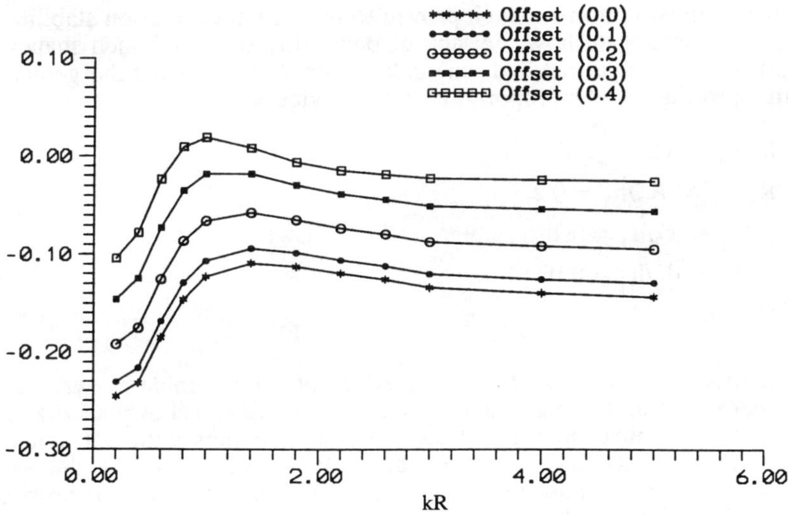

Figure 8c The effect of offset between buoy and plate on the nondimensional vertical added mass for the plate associated with vertical motion of the buoy

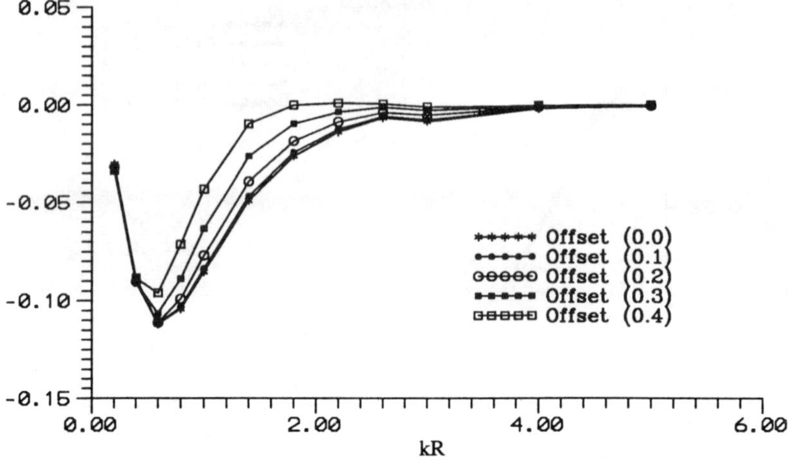

Figure 8d The effect of offset between buoy and plate on the nondimensional vertical potential damping for the plate associated with vertical motion of the buoy

8.2 Buoy and Plate with Conical Bottom

In the Hose Pump field experiments the surface buoy and the submerged plate had conical bottoms, which proved to increase their motion stability. In order to investigate how the lack of possibility to model such shapes in the fast semianalytical method, one calculation was made for the geometry according to Fig. 9. The proportions of the device were

$$h_2/h_1 = 0.3$$
$$R_1/h_1 = R_2/h_1 = 0.2$$
$$r_1/h_1 = r_2/h_1 = 0.1$$
$$d_1/h_1 = d_2/h_1 = 0.0732$$
$$\theta_1 = \theta_2 = 150°$$

The number of panels for a quarter of the problem was 205. Nondimensional added mass and potential damping are shown in Fig. 10a and b for the conical device and the previous cylinders with flat bottoms. ($d1/h1 = d2/h1 = 0.0888$). In both calculations, bodies are of the same volume. The results show that the difference is pronounced but probably tolerable for a first rough optimization of buoy diameter.

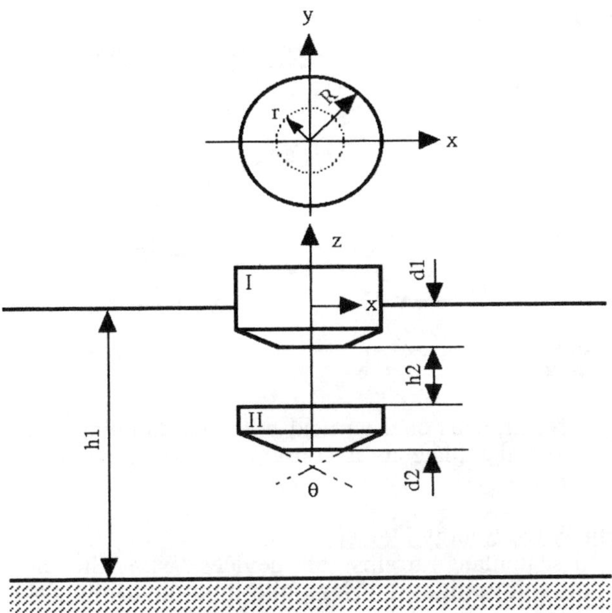

Figure 9 Wave power device with conical bottoms

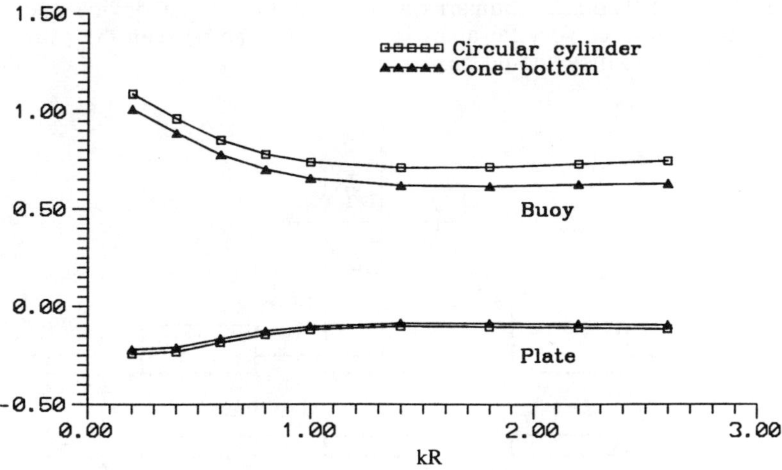

Figure 10a Nondimensional vertical added mass of the buoy and the plate as a consequence of the motion of the buoy

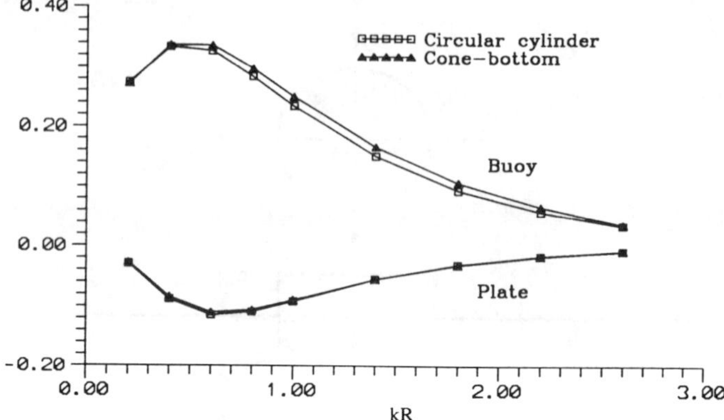

Figure 10b Nondimensional vertical potential damping of the buoy and the plate as a consequence of the motion of the buoy

8.3 Three Adjacent Wave Energy Devices

The possibility to calculate groups of devices with the developed programme was also investigated. Here the hydrodynamic interaction between three devices, i.e. six cylinders is accounted for. See Fig. 11. The same device geometry was used as for the single device with two circular cylinders. But the distance, S, in between the devices were chosen to 500 R. This distance was used to check the consistency of the solution by comparing with the solution for one device.

In Fig. 12a and 12b the comparison between the single device and the group of three devices very far apart is shown. As can be seen the solutions are almost equal as they should be.

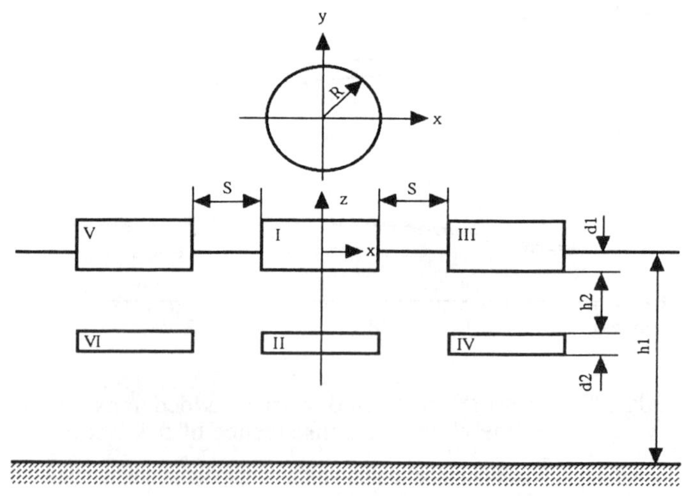

Figure 11 Geometrical properties of the three wave energy devices, with the six bodies I, II VI

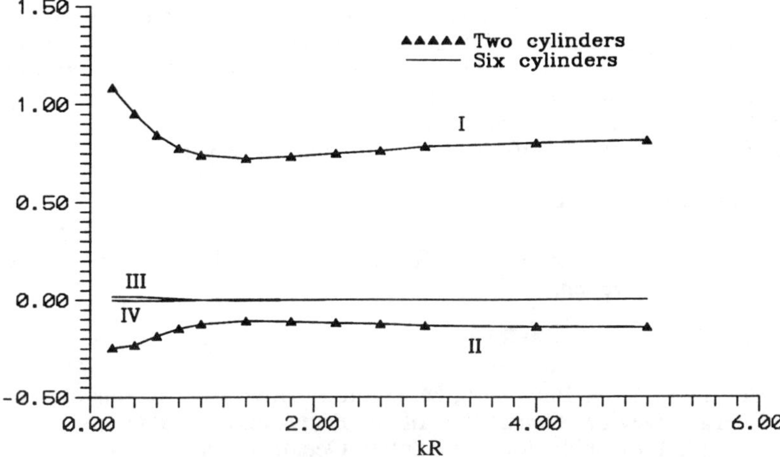

Figure 12a Nondimensional vertical added mass for the six circular cylinders as a consequence of the vertical motion of the middle buoy (S=500R)

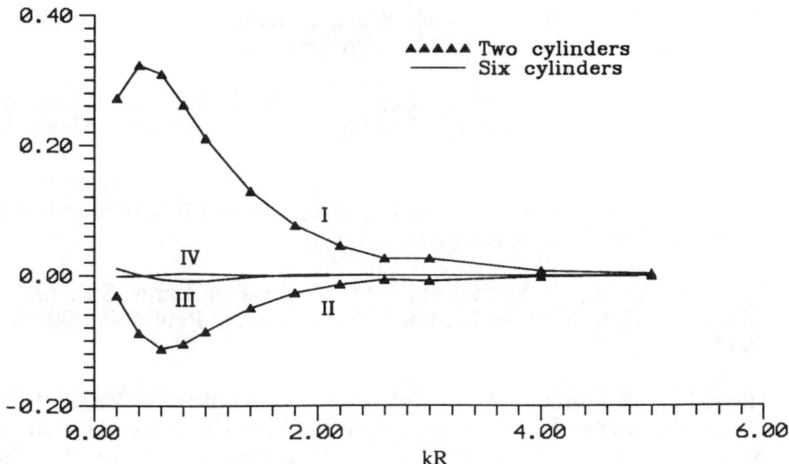

Figure 12b Nondimensional vertical potential damping for the six circular cylinders as a consequence of the vertical motion of the middle buoy (S=500R).

CONCLUSIONS

The developed three-dimensional panel programme is efficient and gives accurate results even for a small number of panels, which was checked by comparing with results for the method of expansion in eigenfunctions. The solution of the radiation problem was also tested against the scatter problem by means of the Haskind relation. The agreement in forces is satisfactory. The development programme seems to be robust and useful.

As concerns the investigation about the possibility to use the fast method of expansion in eigenfunctions for geometries deviating from the simplified geometry used in that method, the present results indicate that this is possible if the deviations are not too great. A correct answer can only be given by putting the present results for hydrodynamic properties into a model for the power take-off. This has not been done yet.

ACKNOWLEDGEMENT

This research was supported financially by the National Energy Administration, Sweden.

REFERENCES

1 Berggren, L., & Johansson, M. Hydrodynamic Coefficients of a Wave Energy Device Consisting of a Buoy and a Submerged Plate. Accepted for publication in Applied Ocean Research, 1992.

2 Berggren, L., & Bergdahl. L. Forces on a Wave Energy Module. Presented at the Third Symposium on Ocean Wave Energy Utilization, Tokyo, Japan 1991.

3 Bj7rkenstam, U., & Lei, X.M. Wave Loading on Multiple Floating Bodies, International Shipbuilding Progress, 1990.

4 Hagerman, G. Wave Energy Resource and Technology Assessment for Coastal North Carolina. SEASUN Power Systems, Alexandria, VA, Aug. 1988.

5 John, F. On the Motions of Floating Bodies I and II. Communications in Pure and Applied Mathematics, Vol. 3, 1950.

6 Faltinsen, O.M., & Michelsen, F.C. Motions of Large Structures in Waves at Zero Froude Number, Nor. Veritas Publ. No. 90, Sept. 1975.

7 Hess, J.L., & Smith, A.M.O. Calculation of Nonlifting Potential Flow about Arbitrary Three-Dimensional Bodies. Douglas Aircraft Co. Report No. E.S. 40622, 1962 (Also in abbreviated form in J. Ship Res., 8, 1964)

8 FINGREEN computer code. Department of Ocean Engineering, MIT Cambridge, USA 1986.

9 Garrison, C.J. in Zienkiewicz, O.C. (Ed.) Numerical methods in offshore engineering. 1978.

A Boundary Element Method Applied to Gas-Liquid Drainage in a Capillary Cavity

D. Lasseux (*), P. Fabrie (*)(**), M. Quintard (*)
(*) L.E.P.T. - ENSAM (UA CNRS), Esplanade des Arts et Métiers, 33405 Talence Cedex, France
(**) UFR de Mathématiques et Informatique (CeReMab - UA CNRS), Université Bordeaux I, 33405 Talence Cedex, France

ABSTRACT
The boundary element method is applied to the Stokes problem with a moving boundary in capillary cavities. The numerical method uses a stress/velocity formulation, which is very suitable for the boundary conditions involved. The displacement of the moving surface is calculated explicitly; smoothing and re-gridding insure stability of the overall scheme. The method allows one to simulate directly the behavior of very thin dynamic films left behind a receding meniscus. The numerical results are compared with both experimental observations and calculations based on a simplified theoretical analysis.

INTRODUCTION

Drainage processes in porous media are of great practical importance in petroleum engineering, a typical problem being gas-oil drainage in the presence of water. Micromodel studies (Chatzis et al., 1988[1]) suggest that these flows are largely influenced by the existence of dynamic films left behind receding menisci. These mechanisms are highly sensitive to pore geometry; however, direct numerical simulations, simplified theoretical results, and experimental studies for simple capillary geometries can help understand the dynamics of such flows.

Simplified solutions, based on lubrication-type approximations, are available for flow in tubes (Bretherton[2]) or in Hele Shaw cells (Park et Homsy[3]). In these configurations, it is possible to derive a simple relationship between the film thickness, e, and the capillary number, Ca, in the following form

$$e/b = 1.337\ Ca^{2/3} \qquad (1)$$

where b is half the distance between the two plates in the Hele Shaw case. In addition to the lubrication-type approximation, this result is based on several simplifications such as: (i) the film thickness is supposed to be constant in a region far from the meniscus, (ii) the Bond number, Bd, is assumed to have negligible effect on the flow. Furthermore, the correlation given by Equation (1) is not valid for large capillary numbers for obvious reasons. Thus, it is very attractive to obtain a full solution of the two-phase flow problem to extend the range of validity of the above correlation.

For more complex geometries, dimensional analysis suggests that such correlations exist. However, numerical simulations are needed to obtain quantitative results. For such moving boundary problems, the boundary element method is very attractive. It can be tested on the simple geometries mentioned above. The Bretherton problem has been studied numerically by Lu[4]. In this paper, we present a numerical solution of the Hele Shaw case. We consider the drainage of a wetting phase (liquid) by a gas phase, the flow taking place between two plates as it is illustrated in Figure 1.

NUMERICAL METHOD

The two-phase flow free boundary problem under consideration is solved numerically using a boundary element technique. This method has proved to be suitable and accurate in many cases of free surface flow as reported in Wrobel and Brebbia[5].

The liquid domain Ω and its boundary $\partial\Omega=S_1\cup S_2\cup S_3$ are illustrated in Figure 1. The liquid phase is limited on each side by two vertical, impermeable, rigid boundaries S_2. The distance between the two vertical planes is equal to 2b. The lower boundary corresponds to a fictitious horizontal plane S_3 in the liquid phase. The upper gas phase is separated from the liquid by the free surface S_1.

The velocity field on S_3 is a Poiseuille flow at a constant flow rate. We assume that the flow pattern in Ω is purely two-dimensional in the plane ($\mathbf{e}_h$, $\mathbf{e}_z$) represented in Figure 1. Furthermore, we assume that: (i) the viscosity and the density of the gas phase are negligible compared to the viscosity and the density of the liquid phase; (ii) the drainage velocity is small enough for the Stokes approximation to be valid in the liquid phase.

As a consequence, the boundary value problem can be written as follows

$$\nabla.\mathbf{S} = \mathbf{0} \tag{2}$$

$$\nabla.\mathbf{v} = 0 \tag{3}$$

in the liquid phase, where $\mathbf{S}$ is the stress tensor, and $\mathbf{v}$ the dimensionless velocity.

The appropriate boundary conditions are

$$\text{(B.C. 1)} \qquad \mathbf{v}\big|_{S_2} = \mathbf{0} \tag{4}$$

$$\text{(B.C. 2)} \qquad \mathbf{v}\big|_{S_3} = (h^2 - 1)\mathbf{e}_z \tag{5}$$

$$\text{(B.C. 3)} \qquad \mathbf{n}.\mathbf{S} = \left(\frac{1/R - Bd\, z}{Ca_0}\right)\mathbf{n} \tag{6}$$

Because of the assumptions we have made, the pressure in the gas phase is assumed to be constant, equal to P_0.

In Equation (2) **S** is the dimensionless stress tensor defined by

$$\mathbf{S} = - p\,\mathbf{I} + \nabla\mathbf{v} + {}^t\nabla\mathbf{v} \qquad (7)$$

in which p is the dimensionless pressure in the liquid phase. The dimensionless velocity is defined from the dimensional corresponding variable, denoted with superscript *, by

$$\mathbf{v} = \mathbf{v}^*/V_0 \qquad (8)$$

where V_0 is the reference velocity on the symmetry axis on S_3. Similarly the dimensionless pressure is defined by

$$p = \frac{p^* - P_0 + \rho\, g\, b\, z}{\mu\, V_0} \qquad (9)$$

where ρ, μ, and g represent the density, the viscosity of the liquid phase and the gravitational acceleration respectively; h and z are the coordinates made dimensionless by b.

In Equation (6), R is the dimensionless radius of curvature at the point of S_1 located at position z; Bd and Ca_0 are respectively the Bond number and the capillary number defined by

$$Bd = \rho\, g\, b^2/\gamma \qquad (10)$$

$$Ca_0 = \mu V_0/\gamma \qquad (11)$$

in which γ is the interfacial tension.

While the liquid phase flows, the free surface S_1, of equation $\Phi(h,z,t) = 0$, evolves following the classical kinematic condition

$$\frac{D\Phi}{Dt} = \frac{\partial\Phi}{\partial t} + \mathbf{v}.\nabla\Phi = 0 \qquad (12)$$

From Equations (2) and (3), one forms the boundary integral equation of the problem (see DaCosta Sequeira[6] and Lasseux[7] for details) which can be written as

$$\int_{\partial\Omega} \mathbf{n}.[\,\mathbf{S}(\mathbf{u}_k(\mathbf{x} - \mathbf{y}), t_k(\mathbf{x} - \mathbf{y})).\mathbf{v}(\mathbf{x}) - \mathbf{S}(\mathbf{v}(\mathbf{x}), p(\mathbf{x})).\mathbf{u}_k(\mathbf{x} - \mathbf{y})\,]\, d\gamma = 0$$

$$k = 1,2 \qquad (13)$$

for any points **x** and **y** on ∂Ω. In this formula, ($\mathbf{u}_1$, $\mathbf{u}_2$) and **t** are the fundamental solution of the elementary two-dimensional Stokes problem

$$-\nabla.[\,(\nabla\mathbf{u}_k(\mathbf{x} - \mathbf{y}) + {}^t\nabla\mathbf{u}_k(\mathbf{x} - \mathbf{y})) - t_k(\mathbf{x} - \mathbf{y})\mathbf{I}\,] = \delta(\mathbf{x} - \mathbf{y})\mathbf{e}_k \quad k=1,2 \qquad (14)$$

where δ(**x** - **y**) is the Dirac distribution at point **y**.

This fundamental solution can be found for example in DaCosta Sequeira[6] or in Ladyzhenskaya[8] and is given by

$$u_{\ell k}(\mathbf{x} - \mathbf{y}) = \frac{1}{4\pi}\left[- \delta_{\ell k} \ln r + \frac{(y_\ell - x_\ell)(y_k - x_k)}{r^2} \right] \qquad (15)$$

$$t_k = \frac{1}{2\pi}\frac{\partial \ln r}{\partial x_k} \qquad (16)$$

with

$$r = |\mathbf{x} - \mathbf{y}| \tag{17}$$

According to the boundary conditions for the problem considered here, equation (13) can be re-written in a more convenient form as

$$\int_{S_2 \cup S_3} \mathbf{n}.[\mathbf{S}(\mathbf{v}(\mathbf{x}), p(\mathbf{x})).\mathbf{u}_k(\mathbf{x} - \mathbf{y})]\, d\gamma - \int_{S_1} \mathbf{n}.[\mathbf{S}(\mathbf{u}_k(\mathbf{x} - \mathbf{y}), \mathbf{t}_k(\mathbf{x} - \mathbf{y})).\mathbf{v}(\mathbf{x})]\, d\gamma =$$

$$\int_{S_2 \cup S_3} \mathbf{n}.[\mathbf{S}(\mathbf{u}_k(\mathbf{x} - \mathbf{y}), \mathbf{t}_k(\mathbf{x} - \mathbf{y})).\mathbf{v}(\mathbf{x})]\, d\gamma - \int_{S_1} \mathbf{n}.[\mathbf{S}(\mathbf{v}(\mathbf{x}), p(\mathbf{x})).\mathbf{u}_k(\mathbf{x} - \mathbf{y})]\, d\gamma$$

$$k = 1,2 \tag{18}$$

At this point, it is seen that the velocity/normal stress formulation is very suitable for the problem under consideration since the boundary conditions appear directly in the above equation. This last equation allows to compute the normal stress **n.S** on $S_2 \cup S_3$, and **v** on the *moving boundary* S_1.

Discretization:

The continuous problem (Equations (2) to (6) and (12)) is now discretized in order to produce a linear system and give the unknown values on $\partial\Omega$ (Huyakorn and Pinder[9]). Our computation has been performed using a *constant boundary elements* method which consists in dividing the boundary $\partial\Omega$ into m straight line segments Γ. On each segment, **n.S** and **v** are assumed to have a constant value, equal to the value at the central node.

The choice for constant boundary elements is mainly justified by the fact that this method gives an accurate approximation of the solution either on the boundary or for points inside the domain far enough from the boundary. This has also been pointed out in other works concerning free surface flows of viscous newtonian fluids (Bush[10]). In addition, the numerical scheme for constant boundary elements is fast and simple to implement, making this method a very attractive one (Bush[10], Lu[4], Sugino and Tosaka[11]).

The first step is to discretize the boundary integral given by Equation (18). This is done by writing

$$\sum_{j=m_1+1}^{m_1+m_2+m_3} \left[(\mathbf{n}_j.\mathbf{S}(\mathbf{v}_j, p_j)) \cdot \int_{\Gamma_j} \mathbf{u}_k(\mathbf{x}_j - \mathbf{y}_i)\, d\gamma_j \right] -$$

$$\sum_{j=1}^{m_1} \left[\int_{\Gamma_j} \mathbf{n}_j.\mathbf{S}(\mathbf{u}_k(\mathbf{x}_j - \mathbf{y}_i), \mathbf{t}_k(\mathbf{x}_j - \mathbf{y}_i))\, d\gamma_j \right] . \mathbf{v}_j =$$

$$\sum_{j=m_1+1}^{m_1+m_2+m_3}\left[\int_{\Gamma_j}\mathbf{n}_j.\mathbf{S}(\mathbf{u}_k(\mathbf{x}_j-\mathbf{y}_i),\mathbf{t}_k(\mathbf{x}_j-\mathbf{y}_i))d\gamma_j\right].\mathbf{v}_j -$$

$$\sum_{j=1}^{m_1}\left[(\mathbf{n}_j.\mathbf{S}(\mathbf{v}_j,p_j)).\int_{\Gamma_j}\mathbf{u}_k(\mathbf{x}_j-\mathbf{y}_i)\,d\gamma_j\right]$$

$$\text{for}\quad k=1,2 \quad\text{and}\quad i=1,m_1+m_2+m_3 \tag{19}$$

In this relation, χ_j means that χ is taken at node j (or on the jth element); $\mathbf{x}_j$ and $\mathbf{y}_i$ denote positions of nodes j and i on $\partial\Omega$; m_ℓ is the number of nodes (or elements) on S_ℓ, ℓ=1,3. Node numbering is such that the node number increases while $\partial\Omega$ is described clockwise and node number one is the first node on S_1 which has a positive h coordinate. We will denote m as the total number of nodes ($m=m_1+m_2+m_3$).

The next step consists in discretizing the boundary conditions given by Equations (4) through (6). According to the method used here, this can be done by writing

$$\text{(B.C. 4)}\quad \mathbf{v}_j = \mathbf{v}|_{\Gamma_j} = \mathbf{0} \qquad j=m_1+1,m_1+m_2 \tag{20}$$

$$\text{(B.C. 5)}\quad \mathbf{v}_j = \mathbf{v}|_{\Gamma_j} = (h_j^2-1)\mathbf{e}_z \qquad j=m_1+m_2+1,m_1+m_2+m_3 \tag{21}$$

$$\text{(B.C. 6)}\quad (\mathbf{n.S})_j = \left[\frac{1/R_j - Bd\,z_j}{Ca_0}\right].\mathbf{n}_j \qquad j=1,m_1 \tag{22}$$

where R_j and $\mathbf{n}_j$ are the radius of curvature of S_1 at node j and the unit outward normal vector of element j respectively.

At this point, one clearly sees that the right hand side of Equation (19) is completely known. In addition, the symmetry of the problem gives a solution symmetric with respect to the z-axis. Therefore the number of unknowns is reduced accordingly.

From Equations (15), (16) and (19), it is clear that the fundamental solution of the elementary two dimensional Stokes problem has singularities at each interpolation node. However, because of the constant boundary elements method retained here, each integral in Equation (19) can be explicitly calculated (see Lasseux[7]), therefore avoiding the use of approximations with quadrature formulae such as those used in Sato et al.[12].

The last discretization step consists of finding a suitable discretized form of Equation (12). Points on the free boundary are moved according to

$$z_j^{n+1} = z_j^n + \delta t\,\mathbf{v}_j^n.\mathbf{e}_z \tag{23}$$

$$h_j^{n+1} = h_j^n + \delta t\, \mathbf{v}_j^n.\mathbf{e}_h \qquad (24)$$

where δt represents the time step, and superscripts n and n+1 denote variables at time steps n and n+1 respectively. Equations (23) and (24) are the approximate form of Equation (12) using an explicit Euler method. This approximation has also been employed successfully for updating the free boundary in Sugino and Tosaka[11] and Sato et al.[12].

In this work, the relations (23) and (24) have been used with the empirical stability condition

$$\delta t < Ca_0/10 \qquad (25)$$

which expresses the fact that the time step is inversely proportional to the normal stress on S_1 (see for example (B.C. 3) in Equation (6)).

Even under the constraint expressed by Equation (25), the updating process described by Equations (23) and (24) generates instabilities which can be observed after only few time steps. This fact has also been noted by Sugino and Tosaka[11]. The reason for this lies in the fact that the element size on S_1 is altered by the updating process. This is due to the velocity gradient along S_1 which causes elements near the symmetry axis to become larger while elements near S_2 become smaller. To remedy this problem, a smoothing and relocation technique has been used. At each time step, the free boundary S_1 is smoothed with local cubic spline functions. These piecewise functions allow one to relocate and eventually add nodes on S_1 in order to keep an approximately constant element size during the whole computation. In addition, as the z coordinate of the lower point on S_1 decreases, it is necessary to increase the vertical size of the domain, i.e., increase the number of nodes m_2 on S_2.

Numerical experiments showed that the above mentioned instabilities are dramatically enhanced when capillary effects are dominant, and this increases the stiffness of the problem while an accurate evaluation of the radius of curvature R of S_1 (see Equation (22)) is required. As the spline functions are only piecewise functions, the evaluation of this last quantity by direct derivation has proved to be unsatisfactory. Hence, the value of R_j is approximated by averaging the distances along the normal line from the jth node to the intersections of the normal lines from the adjacent nodes.

The free boundary is set initially to be the discretized solution of the gravity/capillary two-phase equilibrium. This solution is detailed in Lasseux[7]. Finally, the algorithm is summarized below as

1) The radius of curvature on S_1 is evaluated as described above.
2) Discretized boundary conditions are computed at each node.
3) Integrals in Equation (19) are calculated and the linear system is formed.
4) The linear system is then solved, providing $\mathbf{v}_j$ on S_1 and $(\mathbf{n.S})_j$ on S_2 and S_3. The system is solved by using a QR factorizing method. This method has been chosen on the basis of stability criteria, knowing that the system matrix, which is fully populated and non symmetric, is more ill

conditioned as the discretization is refined. An alternative algorithm for solving this type of system can be found in Cai et al.[13].

5) The free boundary is updated according to Equations (23) (24) and (25).
6) The boundary S_1 is smoothed. Boundaries S_1 and S_2 are re-gridded, keeping the initial element size and vertical extent.

Calculations continue until a significant volume of fluid is displaced. At each time step, precision is checked via the fluid volume balance, which is better than 1% for all our simulations.

Results

Our simulations have been performed for two values of the Bond number (Bd = 0 and Bd = 1.03) and different values of the capillary number. As the total calculation time becomes prohibitive when the capillary number is small (see Equation (25)), our numerical experiments are performed for capillary numbers greater than 2.10^{-3}. In addition, our numerical experiments presented here are limited to the case of perfectly wetting liquid.

In Figure 2 are shown the right hand side shapes of the symmetrical free surface for various times and the two Bond numbers. This figure clearly shows the wetting film, of approximately constant thickness, on the immobile boundary S_2. Furthermore, the important effect of the Bond number is unambiguously outlined as a much thicker film is obtained when gravity forces are set equal to zero. This is confirmed on Figure 3 where our numerical results, in terms of film thickness, are represented as a function of the capillary number and the Bond number. These results suggest that a consistent power law relationship between the film thickness, e, (computed at $z = (z_{m_1} - z_1)/2$) and the capillary number Ca (calculated from the mean velocity V of the first node on S_1 during the overall computation) can be found. As shown in this figure, this kind of relationship is valid for capillary numbers smaller than 0.1. A power law fitted to these numerical results, in this range of capillary numbers, gives

$$e/b = 0.71\ Ca^{2/3} \quad ; \quad Bd = 1.03 \qquad (26)$$

$$e/b = 1.17\ Ca^{2/3} \quad ; \quad Bd = 0 \qquad (27)$$

These results are compared with our theoretical and experimental ones in the following. It should be noticed however that the last equation is in good agreement with the analytical result given by Park and Homsy[3]. In addition, for large values of the capillary number (Ca>0.1), our calculations indicate that the film thickness reaches an asymptotical value of 0.35. This value has also been obtained in several experimental, theoretical and numerical works in similar cases (Reinelt and Saffman[14], Schwartz et al.[15], and Lu[4]).

THEORETICAL RESULTS

The theoretical approach we developed for an approximate analysis of the free boundary problem is summarized below (see Park and Homsy[3] for details). Our calculations are performed in a coordinate system fixed relative to the lower point of the free surface moving downward at a constant velocity V. The free boundary, which is assumed to be symmetric with respect to $\mathbf{e}_z$, is divided into three regions: (1) the region far upstream from the meniscus, called the film developed region, in which capillary forces are small and where the film thickness is assumed to be stationary and equal to e_0; (2) the transition region where all the forces involved in the physical process are of the same order of magnitude, a priori; the film thickness is denoted e in this region; (3) the meniscus region, which is assumed to be identical to the meniscus at equilibrium under capillary and gravity forces in the same cavity (this requires that the capillary number is small).

In region (1) and (2), the lubrication approximation is retained. After the flow rate balance condition is fulfilled between these two regions, and assuming that

$$\mathrm{Ca} = \frac{\mu V}{\gamma} << 1 \tag{28}$$

one can write the approximate differential equation governing the free surface shape away from the meniscus in a dimensionless form as

$$X_{\varepsilon\varepsilon\varepsilon} = \frac{1 - X}{X^3} \tag{29}$$

with

$$X = e / e_0 \tag{30}$$

$$\varepsilon = (3\,\mathrm{Ca})^{1/3}\, z/e_0 \tag{31}$$

The associated boundary conditions are

$$\text{(B.C. 7)} \quad X \rightarrow 1 \quad \text{as} \quad \varepsilon \rightarrow +\infty \tag{32}$$

$$\text{(B.C. 8)} \quad X_\varepsilon \rightarrow 0 \quad \text{as} \quad \varepsilon \rightarrow +\infty \tag{33}$$

$$\text{(B.C. 9)} \quad X_{\varepsilon\varepsilon} \rightarrow 0 \quad \text{as} \quad \varepsilon \rightarrow +\infty \tag{34}$$

The boundary condition (B.C. 8) indicates that the liquid perfectly wets the solid, which is the restriction of our study.

While the flow pattern is complicated in the meniscus region (its description is beyond the scope of this paper), a satisfactory matching condition between the transition region and the meniscus region is given by the required continuity of the curvature (i.e. the capillary pressure) instead of the flow rate balance. In fact, when the liquid perfectly wets the solid, this matching condition is equivalent to the continuity of the second derivative of X.

Hence, it is necessary to investigate the second derivative behavior of the profile in the transition region as the meniscus region is approached, knowing that e is expected to be large compared to e_0 when approaching the meniscus. This requires integrating Equation (29) to determine the limit value of $X_{\varepsilon\varepsilon}$ as X tends to infinity. After returning to the dimensional form, we obtain (with the notation $\Phi = b - e$)

$$|e_{zz}| = |\Phi_{zz}| = 1.338\, \mathrm{Ca}^{2/3} / e_0 \tag{35}$$

while approaching the meniscus region. In addition, from the study of the meniscus at equilibrium ($h = \Phi(z)$), we get the following equation

$$|\Phi_{zz}| = Bd\,\lambda_2/b \qquad (36)$$

at a point $z=b\,\lambda_2$ corresponding to the contact point, $z = 0$ corresponding to the flat interface level outside the plates. Matching $|\Phi_{zz}|$ between Equations (35) and (36) yields

$$e_0 / b = \frac{1.338\,Ca^{2/3}}{Bd\,\lambda_2} \qquad (37)$$

It can be shown (Lasseux[7]) that this important result is valid for any Bond number. Equation (37) gives Equation (1) when Bd=0, and when Bd=1.03 we have

$$e_0 / b = 0.81\,Ca^{2/3} \qquad (38)$$

As can be seen from Equations (26), (27), (1) and (38), numerical results for the film thickness are in good agreement with these analytical predictions. This comparison confirms the consistency of a power law relationship between the dimensionless film thickness and the capillary number ($Ca<0.1$) and the important effect of the Bond number on the physical process.

Our own experimental results are now presented and compared to both of the above mentioned approaches.

COMPARISON WITH EXPERIMENTAL RESULTS

Several drainage experiments of different silicone oils by air have been performed for capillary numbers ranging from 5.10^{-5} to 2.10^{-2} using a vertical Hele Shaw cell (Lasseux and Quintard[16]). These experiments were such that: (i) the cell walls were perfectly wetted by the oils; (ii) the Bond number was constant equal to 1.03 for all the experiments; (iii) the temperature was constant, equal to 20 °C.

The Hele Shaw cell was first saturated with oil to the height Z_0 and then drained at a constant flow rate using a syringe pump until the meniscus reached the position Z_1 ($|Z_0 - Z_1| \approx 150$ b). This stage corresponds to the physical problem studied in this paper. In order to estimate the film thickness, the experiments were subsequently conducted as follows. The bulk drainage was stopped and fluids inside the cell were isolated. Then, the film remaining on the walls drained under gravity and produced a free surface rise until final equilibrium, for which the interface reached a position labelled Z_f. All the positions referred above are those of the lower point of the meniscus on the cell symmetry axis and were detected by photographs. The capillary number was evaluated using the mean velocity of the lower point of the interface, calculated from the positions of this point during the bulk drainage period.

From a volume balance equation, assuming that the film thickness, e, is constant during the bulk drainage, and neglecting the adsorbed film, it can be shown that

$$e / b = \frac{Z_f - Z_1}{Z_0 - Z_1} \qquad (39)$$

In Figure 4 are reported our experimental dimensionless film thickness results versus the capillary number, together with the numerical and analytical predictions. The comparison shows a rather good agreement since the following power law

$$e / b = (0.97 \pm 0.14)\, Ca^{(0.67 \pm 0.02)} \qquad (40)$$

has been fitted on our experimental data.

CONCLUSION

A boundary element method has been successfully developed to describe the Stokes free boundary value flow within two plane parallel plates (Hele Shaw cell). Our numerical code allows to determine the film thickness behind the receding interface during the constant flow rate drainage of a wetting liquid by a gas with a good accuracy. The numerical results are in good agreement with both extended analytical predictions of the film thickness for any value of the Bond number and with experimental data when the capillary number is smaller than 0.1. In this range of capillary number, the consistency of a power law relationship between the film thickness and the capillary number is unambiguously outlined. Furthermore, the important effect of the Bond number on the physical process is confirmed. For large capillary numbers (Ca>1) and any value of the Bond number, a limit film thickness is obtained. This limit value corresponds to that reported in the literature.

Acknowledgement: This work was supported by Institut Français du Pétrole.

REFERENCES

1. Chatzis, I., Kantzas, A. and Dullien, F.A.L. On the investigation of gravity assisted inert gas injection using micromodels, long Berea cores and computer assisted tomography, SPE paper 18284, *63rd SPE Annual Technical Conf. and Exhibition*, Houston, 1988.
2. Bretherton, F.P. The motion of long bubbles in tubes, J. Fluid Mech. 10, 166-188, 1961.
3. Park, C.W. and Homsy, G.M. Two-phase displacement in Hele Shaw cells: Theory, J. Fluid Mech. 139, 291-308, 1984.
4. Lu, W.Q. 'Boundary Element Analysis of Free Surface Problems of Axisymmetric Taylor Bubbles', Boundary Elements XII, Edited by M. Tanaka, C.A. Brebbia and T. Honna, Springer-Verlag, New York. Pro. 12th Int. Conf. on Boundary Elements in Engineering, Sapporo, September 1990.

5. Wrobel, L.C. and Brebbia, C.A. (Ed.) 'Fluid Flow', Section 4, *Free Surface Flow,* Pro. 1st Int. Conf. on Computational Modelling of Free and Moving Boundary Problems, Southampton, CML Publications, Southampton, Boston, 1991.
6. DaCosta Sequeira, A. Couplage entre la méthode des éléments finis et la méthode des équations intégrales: application au problème de Stokes extérieur dans le plan, Thèse Doc. Université de Paris VI, 1981.
7. Lasseux, D. Caractérisation expérimentale, analytique et numérique d'un film dynamique lors du drainage d'un capillaire, Thèse Doc. Université de Bordeaux I.
8. Ladyzhenskaya, O.A. *The Mathematical Theory of Viscous Incompressible Flow*, Gordon & Breach, New York, 1969.
9. Huyakorn, P.S. and Pinder, G.F. *Computational Methods in Subsurface Flow*, Academic Press, 1983.
10. Bush, M.B. 'Stratified Flows of Newtonian Viscous Liquids', Boundary Elements XII, Edited by M. Tanaka, C.A. Brebbia and T. Honna, Springer-Verlag, New York. Pro. 12th Int. Conf. on Boundary Elements in Engineering, Sapporo, September 1990.
11. Sugino, R. and Tosaka, N. 'Boundary Element Analysis of Unsteady Nonlinear Surface Wave on Water', Boundary Elements XII, Edited by M. Tanaka, C.A. Brebbia and T. Honna, Springer-Verlag, New York. Pro. 12th Int. Conf. on Boundary Elements in Engineering, Sapporo, September 1990.
12. Sato, K., Tomita, Y., and Shima, A. 'Numerical Analysis of the Behavior of a Cavitation Bubble near a Vibrating Rigid Wall by the Boundary Integral Method', Boundary Elements XII, Edited by M. Tanaka, C.A. Brebbia and T. Honna, Springer-Verlag, New York. Pro. 12th Int. Conf. on Boundary Elements in Engineering, Sapporo, September 1990.
13. Cai, R.Y., Zeng, Z.J., and Chen F. 'A Partitioning Solution of Non-Symmetrical Fully Populated Matrix System in the Boundary Element Method and its Subroutines', Boundary Elements XII, Edited by M. Tanaka, C.A. Brebbia and T. Honna, Springer-Verlag, New York. Pro. 12th Int. Conf. on Boundary Elements in Engineering, Sapporo, September 1990.
14. Reinelt, D.A., and Saffman, P.G. The penetration of a finger into a viscous fluid in a channel and tube, SIAM J. Sci. Stat. Comput. 6 (3), 542-561, 1985.
15. Schwartz, L.W., Princen, H.M., and Kiss, A.D. On the motion of bubbles in capillary tubes, J. Fluid Mech. 172, 259-275, 1986.
16. Lasseux, D., and Quintard, M. Epaisseur d'un film dynamique derrière un ménisque récessif, to be published in C. R. Acad. Sc. Paris, Série II, 1991.

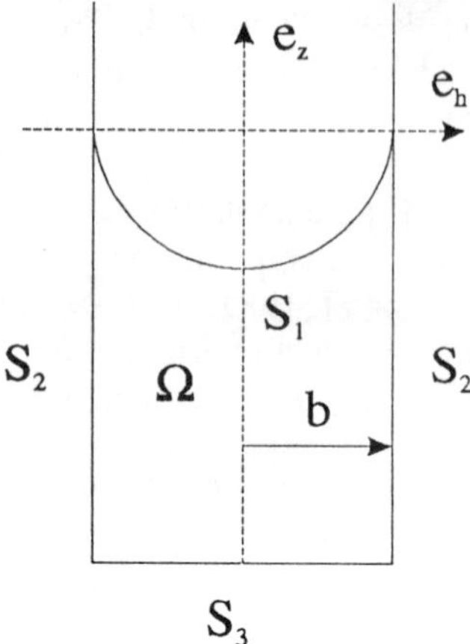

Figure 1. Geometry.

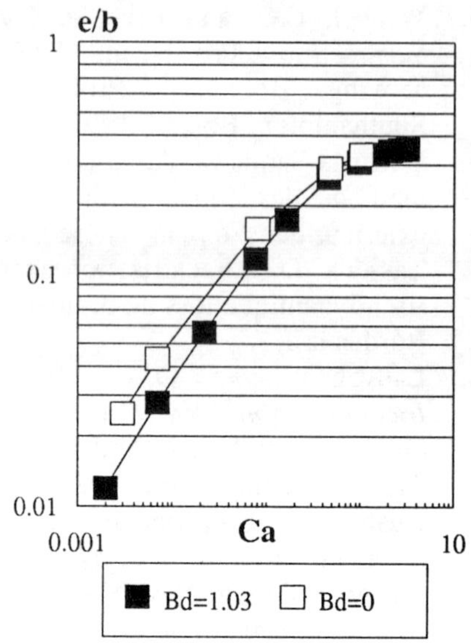

Figure 3. Numerical results.

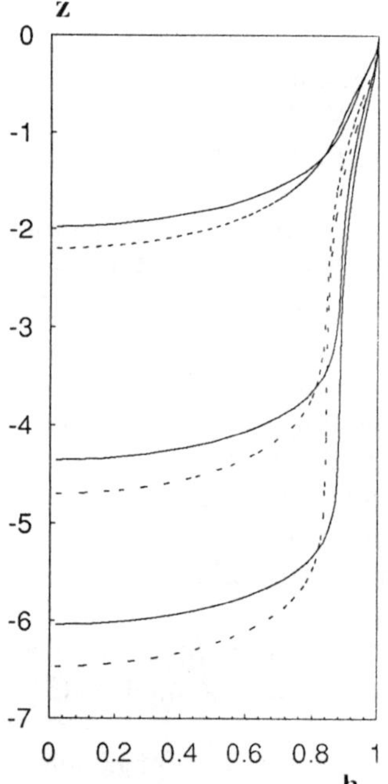

Figure 2. Position of the interface at t_1=1.44 , t_2=4.48 , t_3=7 for Ca_0=.1. The solid line corresponds to Bd=1.03, the discontinuous line corresponds to Bd=0.

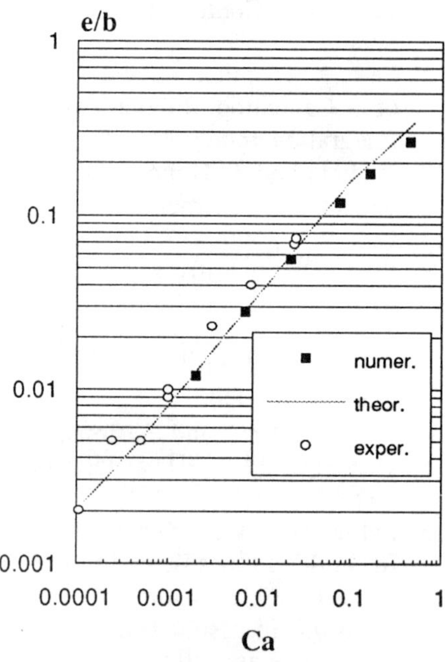

Figure 4. Comparison of the results obtained for Bd=1.03.

Application of the Boundary Element Method to Moving Boundary Problems Arising During Non-Aqueous Phase Liquid (NAPL) Migration in Soils

D.V. Doshi, D.D. Reible

Department of Chemical Engineering, Louisiana State University, Baton Rouge, LA 70803, U.S.A.

ABSTRACT

In evaluating risks associated with chemical spills on the ground surface or leaks from underground storage tanks, it is required to know the extent and degree of contamination in the subsurface. Under certain conditions the contamination front can be treated as sharp and quasi-steady during both initial migration away from a spill and during removal via in-situ extractive processes. The boundary element method (BEM) is a viable choice of numerical method of solution in such situations where only the movement of the boundary is required. BEM is used to solve the set of partial differential equations governing the subsurface flow of NAPL bound by a sharp surface or interface in the vertical plane. This is followed by moving the free boundary using Darcy's law. Unlike other methods the node movement is not restricted to the vertical direction. In order to obtain a high accuracy solution a node redistribution algorithm developed by Carey and Kennon [9] was implemented. It redistributed the nodes according to curvature of the free boundary while retaining the shape. Aquifers with homogeneous as well as piecewise homogeneous properties are considered. Results are presented for simulation of napl infiltration and spreading on an impermeable surface that agree well with laboratory infiltration experiments. Model parameters were obtained by separate measurements and were not fitted to the experimental data.

INTRODUCTION

Subsurface spills of nonaqueous phase liquids (NAPLs) lead to a three phase (air, water, NAPL) transport problem in a fourth essentially immobile media (soil). In general, saturation levels of each of the three fluid phases are variable throughout a transition zone that represents the dispersed front of infiltrating fluid. A variable saturation model perhaps best describes the behavior of the transition zone; however, the equations are difficult to solve and the dispersive parameters are difficult to measure. Many subsurface spill situations, however, develop distinct fronts that move through the reservoir and tend to be self-sharpening. This tends to occur with larger volume spills in that the relative permeability of the NAPL in a fully saturated zone near the spill source tends to be much larger than the relative permeability of the NAPL at a low saturation front, leading to convergence at the front. Saturation changes occur over relatively small distances so that the width of the transition zone is small compared to the aquifer thickness. In such cases the sharp interface approximation may be introduced. Numerical techniques are often used to obtain solutions for these problems.

Hochmuth and Sunada [1] and Weaver [2] developed simplified models based on the assumption of sharp interface between the non-aqueous and the aqueous or the air phase. The earlier model is limited to spreading of the organic phase on a water table, where as the latter is limited to NAPL movement in the unsaturated zone.

In this paper the boundary element method is used to solve the set of partial differential equations describing the subsurface flow of a NAPL bound by a sharp surface or interface in the vertical plane. The assumption of a sharp front was verified for the conditions of the simulation by gamma-ray attenuation measurements in column and two-dimensional soil flume experiments of Reible et al [3]. Because of the speed of propagation of pressure transients is usually far greater than the speed of the fluid in porous media, the flow is treated as quasi-steady. The NAPL migration front is moved at each time step according to fluxes calculated on the boundary assuming a steady pressure field. The validity of this moving boundary model is tested against data obtained in the laboratory experiments.

MATHEMATICAL FORMULATION

The vadose zone infiltration of the NAPL is controlled by the interaction of four phases; bulk NAPL, residual water, air and soil. The basis of the mathematical description of multiphase fluid flow in porous media is the conservation equations for mass and momentum of each phase. Derivations of general multiphase flow models may be found in work by Bear [4]. In most models of non-aqueous phase infiltration the porous medium is assumed rigid and the air phase movement is neglected. The generalized form of Darcy's equation to multiphase flow may be used as simplified momentum balances (Faust[5]). The subsurface movement of water and NAPL phases is governed by

$$\epsilon\frac{\partial S_i}{\partial t}=\nabla(K_i\nabla h_i) \qquad i= \textit{water, napl} \tag{1}$$

where ε is the porosity, S_i is the saturation of phase i, t is time, K_i is the fluid conductivity, h_i is the total head. K_i is defined by

$$K_i=\kappa k_{ri} g\frac{\rho_i}{\mu_i} \tag{2}$$

where κ is the intrinsic permeability of the soil, k_{ri} is the relative permeability of phase i, g is the gravitational acceleration, ρ_i is the density and μ_i is the dynamic viscosity. The relative permeabilities are functions of the fluid saturations. h_i is the total head defined by

$$h_i=\frac{p_i}{g\rho_i}-z \tag{3}$$

where p_i is the phase pressure and z is the vertical coordinate positive *downwards*. Equations (1 to 3) are the starting point for all models of bulk non-aqueous phase movement.

In general, the solution of Equations (1 to 3) requires coupling of the water and NAPL flow equations. In the vadose zone, however, the water saturation is often limited to water tightly held by capillary forces in the typically water-wetted soil. The experiments of Reible et al. [3] indicated that this water was not displaced during the non-aqueous phase infiltration. Thus only the NAPL flow equation is needed to be solved. In addition due to the self-sharpening nature of the fronts under consideration, the NAPL saturation level and thus the relative permeability of a NAPL phase remain constant behind the front. Thus the governing equation (1) reduces to

$$\nabla(K_n\nabla h_n)=0 \tag{4}$$

Although time does not appear explicitly in this equation, potentials are a function of time, since as the front moves, the distribution of fluids in space changes. To be able to solve the problem, we need to define a front function and move it through the domain. In other words if x_f is a vector representing a point on the front at time level t_n, then the coordinate at time t_{n+1} is given as:

$$x_f(t_{n+1}) = x_f(t_n) + v_f \Delta t \qquad \textbf{(5)}$$

where v_f is the front velocity vector and $\Delta t = t_{n+1} - t_n$. At a given location of the front, the flow is treated as steady, due to the assumption of incompressible fluids. When the front moves, a new distribution of fluids is achieved and a new quasi-steady pressure or head field is assumed.

Equation (4) is solved using a Boundary Element Method for two-dimensional boundary value problems (Brebbia et al [6]). The application of Green's theorem and boundary discretization gives:

$$C_\Omega(\phi_\Omega) + G(q_\Omega) - H(\phi_\Omega) = 0 \qquad \textbf{(6)}$$

where C and the matrices G and H depend on the geometry of the system.

Kemblowski [7] used the BEM with implicit scheme to solve for the motion of salt-water interface. The implicit scheme was used because the explicit scheme may cause instability of the numerical solution. Similar problems were identified during the early stages of this work. The implicit method requires computation of fluxes twice for each time step. During simulations it was observed that the fluxes computed at consecutive small time steps were almost the same implying that the cause of front instability was the geometry of the front. Thus the front was advanced in small time steps so that at each movement, the normal to secant is truly normal to the curve at a given point. The calculation of the pressure or head field and thus the NAPL infiltration velocity was calculated on a much larger time step. Advancing the front in small time steps had the advantage of eliminating the need for an implicit scheme of flux calculation and repeated node redistribution. This resulted in approximately 50% reduction of computer time compared to an implicit scheme. Kemblowski [7], and earlier Liggett [8], had restricted the movement of the material surface nodes to only the vertical direction, whereas this work does not put any such limitation.

Adaptive Grid Scheme

Experimentation with node spacing showed that placing more nodes in a high flux or high curvature area improved the accuracy of the prediction. An adaptive grid is then useful where one may fix the number of grid points but adaptively move the grid points so that the solution behavior in layers and near singularities is well modelled. Carey and Kennon [9] extended their earlier work on grading functions for one-dimensional problems to the boundary counter and mesh in a boundary solution method. Important features of which were the determination of a grading function for the curved contour, solution of the boundary equations with mesh adjustment and preservation of the geometry as the mesh is redistributed. A summary of the application of their approach to this problem is given below.

Given a continuous function construct an appropriate mesh for interpolating. Define a grading function ξ(s) such that,

$$\xi_i(s) = \frac{i}{N} \qquad i = 0, 1, \ldots, N \tag{7}$$

For a linearly interpolated boundary an appropriate grading function was found to be (Carey and Kennon [9]),

$$\xi(s) = \frac{\int_0^s [(x'')^2 + (y'')^2]^{0.2}\, ds}{\int_0^S [(x'')^2 + (y'')^2]^{0.2}\, ds} \tag{8}$$

s_i was adjusted using Newton-Raphson iterations to satisfy the previous equation for each node.

Heterogeneous Domain

Consider a problem domain defined over a region which is piecewise homogeneous. The governing equation for this type of problem is

$$\nabla^2 \phi = 0 \tag{9}$$

in each region. Therefore, the boundary element analysis requires the separate discretization of each region. At the interface between the different materials, the elements describing each region share common nodal points. At these nodes the front must satisfy two conditions which mathematically can be stated as

$$q_I^{(1)} = -q_I^{(2)} = q_I \tag{10}$$

$$p_1 = p_2 + p_{c12} \tag{11}$$

$$\rho^{(1)} g (\phi_I^{(1)} + z_I) = \rho^{(2)} g (\phi_I^{(2)} + z_I) + p_{c12} = \phi_I \tag{12}$$

Let

$$h_{c12} = \frac{p_{c12}}{\rho^{(2)} g} \tag{13}$$

where I represent the interface and 1 and 2 represent the regions. The first condition requires that the flux of material out of region 1 must equal the flux into region 2. The negative sign in equation (10) represents the direction of the outward normal on the interface for each region. Second, the pressure difference must equal the capillary pressure that marks the interface between the two fluids. For constant saturations of fluids 1 and 2 we have constant capillary pressure ($p_{c12} = p_2 - p_1$) at the interface.

A system of equations for each region is obtained. These systems of equations are combined by applying the conditions presented in equations (10) and (12) and representing the fictitious interfacial potential and flux by Φ_I and q_I, respectively.

For region 1 equation (6), after absorbing C in matrix H, in the matrix form may be written as

$$\begin{bmatrix} H^{(1)} & H_I^{(1)} \end{bmatrix} \begin{Bmatrix} \phi^{(1)} \\ \phi_I^{(1)} \end{Bmatrix} = \begin{bmatrix} G^{(1)} & G_I^{(1)} \end{bmatrix} \begin{Bmatrix} q^{(1)} \\ q_I^{(1)} \end{Bmatrix} \tag{14}$$

and after substituting capillary pressure condition may be rearranged as

$$\begin{bmatrix} H^{(1)} & \frac{1}{\rho^{(1)}} H_I^{(1)} \end{bmatrix} \begin{Bmatrix} \phi^{(1)} \\ \phi_I^{(1)} \end{Bmatrix} = \begin{bmatrix} G^{(1)} & G_I^{(1)} \end{bmatrix} \begin{Bmatrix} q^{(1)} \\ q_I \end{Bmatrix} + H_I^{(1)} z_I \tag{15}$$

Similarly for the zone 2,

$$\begin{bmatrix} H^{(2)} & \frac{1}{\rho^{(2)}} H_I^{(2)} \end{bmatrix} \begin{Bmatrix} \phi^{(2)} \\ \phi_I^{(2)} \end{Bmatrix} = \begin{bmatrix} G^{(2)} & G_I^{(2)} \end{bmatrix} \begin{Bmatrix} q^{(2)} \\ q_I \end{Bmatrix} + H_I^{(2)} (z_I + h_{c12}) \tag{16}$$

Combining equations (15) and (16) gives

$$\begin{bmatrix} H^{(1)} & \frac{1}{\rho^{(1)}} H_I^{(1)} & -G_I^{(1)} & 0 \\ 0 & \frac{1}{\rho^{(2)}} H_I^{(2)} & -(-G_I^{(2)}) & 0 \end{bmatrix} \begin{Bmatrix} \phi^{(1)} \\ \phi_I \\ q_I \\ \phi^{(2)} \end{Bmatrix} = \begin{bmatrix} G^{(1)} & 0 \\ 0 & G^{(2)} \end{bmatrix} \begin{Bmatrix} q^{(1)} \\ q^{(2)} \end{Bmatrix} + \begin{bmatrix} H_I^{(1)} & 0 \\ 0 & H_I^{(2)} \end{bmatrix} \begin{Bmatrix} z_I \\ z_I + h_{c12} \end{Bmatrix} \tag{17}$$

Solution of this system of equations directly provides the potential and fluxes which then is used to move the free boundary.

EXAMPLE PROBLEMS

A continuous source of Automatic Transmission Fluid (ATF, ρ=0.87 g/cc, ν=20 cSt) at the surface of a sandy soil (K_o= 0.126 cm/min, ϵ=0.4, θ_{wr}=0.1) is simulated in the following examples for various geological conditions. A ponded depth of 5 cm. is maintained at the soil surface. The experimental front at 1.5 min. was used as the initial condition.

Homogeneous Medium with Impermeable Bottom

Contaminant migration was simulated in a homogeneous aquifer having a horizontal impermeable stratum 75 cm. below the contamination source. Tangential movement of nodes was included in the simulation. Thus on the impermeable surface tangential movement of the nodes was possible. At each time step the front was moved in 5 to 50 substeps to prevent geometrical distortion. Larger substeps were especially useful during contaminant spread on the impermeable surface. Time step and its subdivisions could be changed interactively during the simulation. It was also possible to stop the simulation at any time make any changes in data and restart. This facility of cold start allowed considerable saving of time and effort. Nodes were added to the moving boundary during simulation due to its increase in size. At the final time step shown 176 nodes were used whereas only 55 nodes were needed at the initiation of the simulation.

Figure 1 shows simulated ATF fronts at various times. The plume width remained unchanged until the front reached the impermeable surface and began spreading. Use of adaptive grid and time step subdivision enabled the use of large time steps. During napl migration while away from the impermeable surface a 15 min. time step could be used. During its spread on the bottom 5 min. time stepping was possible. This further increased computational saving because most of the computational effort goes into setting up the system of nodal equations and solving them.

Heterogeneous formations- lenses

It is very common to find lenses of low permeability, e.g. clay, in the subsurface. Their effect on the migration of napl plume was studied here. A simple case of a rectangular lens (15 cm x 15 cm) placed under the contaminant source was considered. For simplicity the lens was assumed to be relatively impermeable. The model was solved at the lens boundary only when necessary, that is after the NAPL front had arrived at the lens. As expected the napl ponded on the clay lens and then migrated sideways to the edge of the lens and then again began moving downwards. In figure 2 contaminant fronts are shown at various times. An impermeable stratum was placed some distance below the clay lens. Simulated fronts show that the plume begins to merge below the lens.

The behavior predicted by the model in both simulations agree qualitatively and quantitatively (within the uncertainty of the model parameters) with the experiments conducted under the same conditions. The comparison of the experiments and model simulations are not discussed in detail here.

CONCLUSIONS

Two-dimensional unsaturated infiltration into a shallow water table, residually water saturated aquifer was modeled using laboratory based hydraulic model parameters and sharp front assumption. BEM is a very useful tool for solving such moving boundary problems. Its economy is enhanced with the use of adaptive grid which increases accuracy while allowing simulation with smaller number of nodes. Its application to the current problem was also enhanced by updating the geometry of the migrating front on a much smaller time scale than the calculation of the pressure or head field and the resulting front velocities.

REFERENCES

1. Hochmuth,D.P. and Sunada, D.K. 'Ground-Water Model of Two-Phase Immiscible Flow in Coarse Material' *Ground Water*, *23*, 617-626, 1985
2. Weaver, J.W. *Kinematic Modeling of Multiphase Surface Transport* Ph.D. dissertation, U. Texas, Austin, 1988
3. Reible, D.D., Illangasekare, T.H., Doshi, D.V. and Malhiet, M.E. 'Infiltration of Immiscible Contaminants in the Unsaturated Zone' *Ground Water*, *28*,685-692, 1990
4. Bear, J. *Dynamics of Fluids in Porous Media*, Elsevier, N Y, 1972
5. Faust, C.R. 'Transport of Immiscible Fluids Within and Below the Unsaturated Zone: A Numerical Model' *Water Resources Research 21*, 587-596, 1985
6. Brebbia, C.A., Telles, J.C.F. and Wrobel, L.C. *Boundary Element Techniques* Springer- Verlag, Berlin and New York, 1984
7. Kemblowski,M. 'Salt-Water-Freshwater Transient Upconing - An Implicit Boundary Element Solution' *J. of Hydrology*, *78*, 35-47, 1985
8. Liggett, J.A. 'Locations of Free Surfaces in Porous Media' *J. Hyd. Div., ASCE*, *103*, 353-365, 1977
9. Carey,G.F. and Kennon,S. 'Adaptive Mesh Redistribution for a Boundary Element (Panel) Method' *Int. J. Num. Methods Eng.*, *24*, 2315-2325, 1987

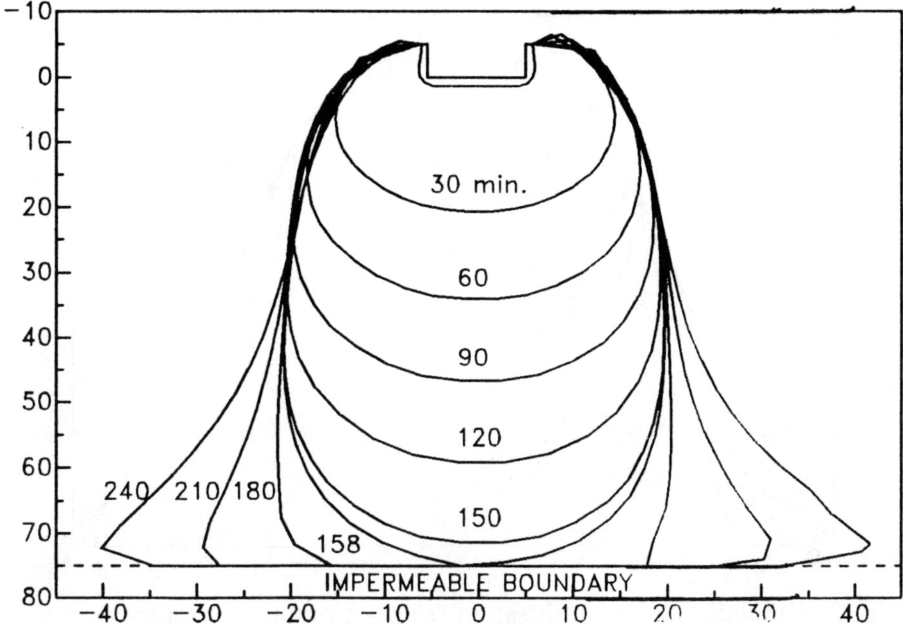

Fig. 1 Simulated NAPL Infiltration from a Continuous Source in Homogeneous Medium with Impermeable Boundary

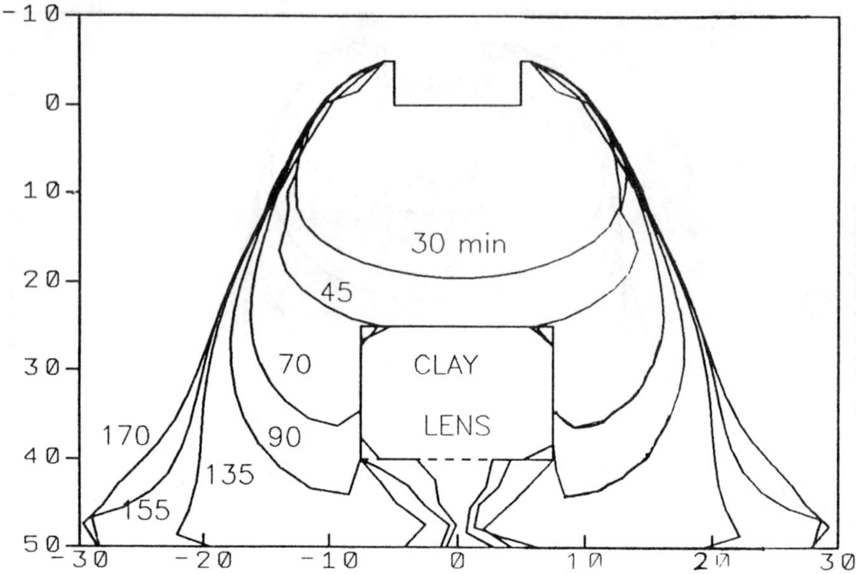

Fig. 2 Simulated NAPL Infiltration from a Continuous Source in Heterogeneous Porous Medium

Free Surface Flows Induced by a Submerged Source or Sink from a Three-Layer Fluid with Stagnation and Cusp Points

X. Wen

Department of Applied Mathematical Studies, University of Leeds, Leeds LS2 9IT, U.K.

ABSTRACT

The free surfaces produced by a submerged line sink or source from a region containing three homogeneous layers of fluid at different densities is considered. The flow is assumed to be steady, ideal and irrotational and solutions are obtained using the boundary integral method. Solutions are obtained in which there is a cusp point on one free surface and a stagnation point on the other. The Froude number on the free surface with a cusp point is infinite whilst on the free surface with a stagnation point the Froude number is small. A numerical method is developed in order to obtain the profiles of the free surfaces and the position of the sink or source. It is found that when the Froude number is fixed then there is a range of positions of the sink or source for which there is a solution and outside this range there are no solutions.

INTRODUCTION

Selective withdrawal of fluid from a reservoir has received a great deal of attention over the last 20 years[1]. It has a wide range of applications as a management technique for the supply of water with desired water quality properties. The steady, two-dimensional free surface flows which are produced by a submerged source, or sink, with gravity as the restoring force has been the subject of many papers which shed deeper insight into the physical principles in order to obviate the need for mechanical devices such as pumps or bubblers.

In a two-layer fluid the situation in which a stagnation point exists directly above the source, or sink, has been considered by Craya [2] and Yih [3], Peregrine[4], Vanden Broeck, Schwartz and Tuck [5], Hocking and Forbes [6], and Mekias and Vanden-Broeck [7]. Further, there have been several papers dealing with flows in which a cusp point exists on the free surface directly above the source, or sink, by Tuck and Vanden-Broeck [8], Hocking [9] Colling [10], Vanden Broeck and Keller [11], Hocking [12], King and Bloor [13], and Hocking [14].

There are many examples of where the flows described above occur in practical situations, for example, the withdrawal of a warm liquid (water) which is above a cooler and more dense liquid (water) and for which there is air above the warm liquid. The warm water is bounded by two interfaces and it is withdrawn from a sink. We are interested in determining the behaviour of the flow so that we may assign the position of the sink so that only warm liquid is withdrawn. For the three-layer fluid Wen and Ingham [15] found the solutions with cusp point solutions on both free surfaces for supercritical flows. Wen and Ingham [16] also considered another type of solution with stagnation points on both free surface in the special case in which the the Froude numbers on both of the free surfaces are equal. In this paper we consider the situation in which there is a cusp point on one of the free surfaces and a stagnation point on the other free surface. For the purposes of this paper we will assume that the fluid is injected into the middle layer (see Fig. 1).

FORMULATION

Because of the quadratic dependence of the velocity on the free surface height then the direction of a potential flow may be reversed and the results for a submerged source may be obtained from the solution for a submerged sink by merely a change of sign. For convenience, we therefore consider here the flow due to a source.

We consider three layers of fluid each of which have different constant densities ρ_1, ρ_2 and ρ_3 ($\rho_1 < \rho_2 < \rho_3$). The light and heavy fluids occupy a semi-infinite region of space and are separated by an infinite strip of fluid of height H and density ρ_2 in the middle layer, see Fig. 1. A source of strength 2q is now situated at the point S which is situated at a height H_s above the equilibrium level of the top of the lower fluid. We assume that there is a cusp point on one free surface and a stagnation point on the other free surface (see Fig. 1). The flow is assumed to be steady, inviscid, incompressible and irrotational and moves under the action of gravity.

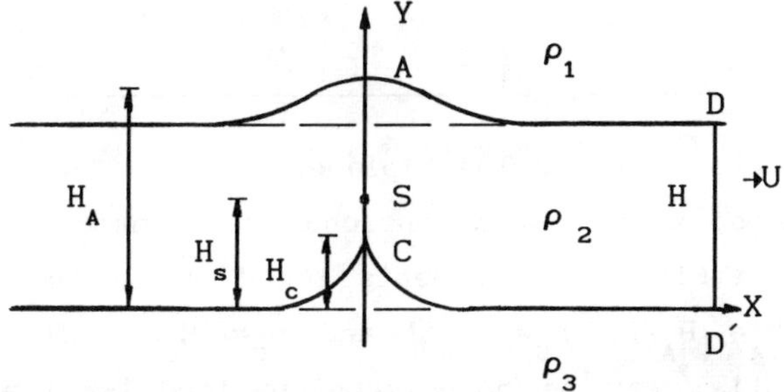

Fig. 1 Flows when there is cusp point on one free surface and a stagnation point on the other surface.

A coordinate system $Z=X+iY$ is introduced with the X axis along the undisturbed lower interface and the Y axis through the source S. In Fig. 1, H_A and H_C are the heights of the stagnation point A and the cusp point C, respectively. Far from the source the flow in the middle layer of fluid is uniform and has speed U away from the source. Since the flow is symmetrical we need only consider the right hand half-plane $X > 0$. The complex velocity potential is defined by $W = \Phi+i\Psi$, where Φ is the velocity potential and Ψ is the streamfunction. Without any loss of generality we choose $\phi = 0$ at the point A and $\Psi = 0$ to be SCD′ and $\Psi = Uh$ on SAD, see Fig. 1.

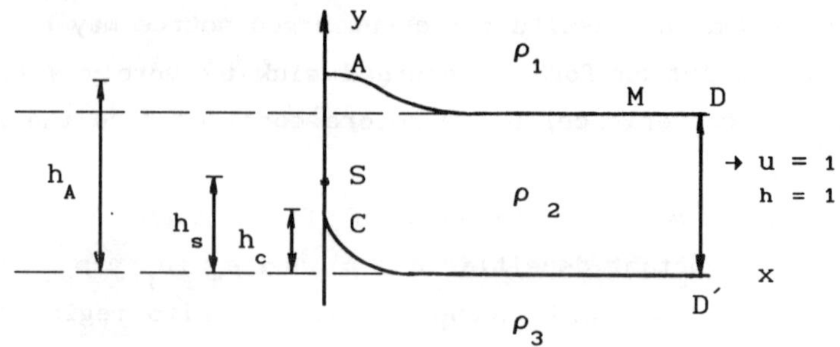

Fig. 2a The physical z plane.

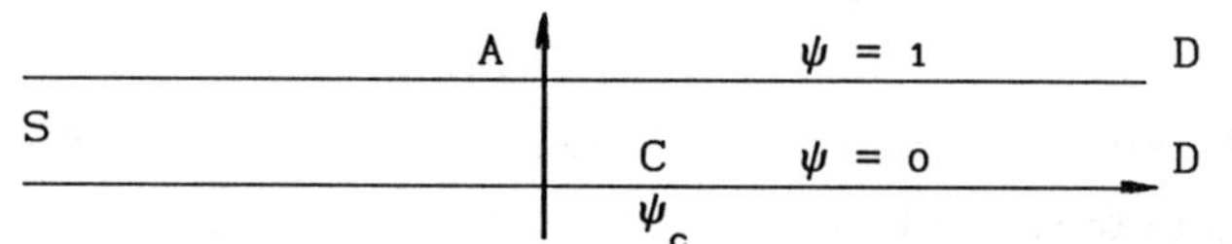

Fig. 2b The W-plane.

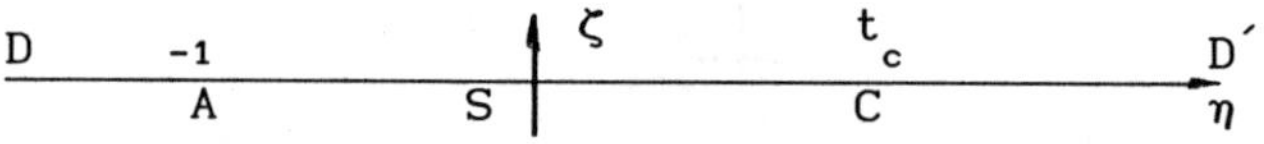

D -1 A S ζ t_c C D´ η

Fig. 2c The t-plane.

We now choose a new non-dimensional coordinate $z = x+iy = (X + iY)/H$, and Fig. 2a represents the flow in the physical plane where $h_A = H_A/H$, $h_S = H_S/H$ and $h_C = H_C/H$. The complex velocity potential may be non-dimensionalised by $w = \phi+i\psi = (\Phi+i\Psi)/UH$ and the strip in Fig. 2b represents the flow in the w-plane and all speeds have now been non-dimensionalised with respect to U.

Along the free surfaces AD and CD′ we apply Bernoulli's equation in the form

$$\frac{u^2}{2} Fr1^2 + y = \frac{Fr1^2}{2} + 1 \qquad \text{on AD} \tag{1}$$

$$\frac{u^2}{2} Fr2^2 - y = \frac{Fr2^2}{2} \qquad \text{on CD}' \tag{2}$$

where the Froude numbers are defined by $Fr1 = U/\sqrt{g_1 H}$, $Fr2 = U/\sqrt{g_2 H}$, where $g_1 = g(\rho_1 - \rho_2)/\rho_2$ and $g_2 = g(\rho_3 - \rho_2)/\rho_2$. We next consider the complex velocity

$$\frac{dw}{dz} = ue^{-i\theta} \tag{3}$$

in which u is the non-dimensionalised fluid speed and θ is the angle that the velocity vector on the free surface makes with the positive x axis. Then the logarithm of the complex velocity is

$$\Omega = \mathrm{Ln}\left(\frac{dw}{dz}\right) = \tau - i\theta \tag{4}$$

where $\tau = \mathrm{Ln}\ u$, and τ, $\frac{dw}{dz}$ and Ω are analytical functions in the strip of the w-plane, see Fig. 2.

We now map the strip in the complex potential plane onto the upper half-plane of the auxiliary t-plane by the transformation

$$t = e^{\pi w} \qquad \text{or} \qquad w = \ln\left(\frac{t}{\pi}\right) \tag{5}$$

The boundary conditions on the real η axis of the t-plane are

$$\begin{aligned}
&\mathrm{Im}\Omega(\eta) = -\beta(\eta) && -\infty < \eta < -1\\
&\mathrm{Im}\Omega(\eta) = -\pi/2 && -1 < \eta < 0\\
&\mathrm{Im}\Omega(\eta) = \pi/2 && 0 < \eta < t_c\\
&\mathrm{Re}\Omega(\eta) = \tau(\eta) = \frac{1}{2}\,\mathrm{Ln}\left(1 + \frac{2y(\eta)}{F_{r2}^2}\right) && t_c < \eta < \infty
\end{aligned} \tag{6}$$

The problem now reduces to a mixed boundary-value problem in the upper half t-plane. By referring to the general solution of the Riemann-Hilbert problem [17] we obtain the solution of Ω in the form

$$\Omega(t) = \frac{\sqrt{t_c - t}}{\pi}\left(\int_{-\infty}^{-1} \frac{-\beta(\eta)}{\sqrt{t_c - \eta}\,(\eta - t)}\,d\eta - \frac{\pi}{2}\int_{-1}^{0} \frac{1}{\sqrt{t_c - \eta}\,(\eta - t)}\,d\eta \right.$$
$$\left. + \frac{\pi}{2}\int_{0}^{t_c} \frac{1}{\sqrt{t_c - \eta}\,(\eta - t)}\,d\eta - \int_{t_c}^{\infty} \frac{\tau(\eta)}{\sqrt{\eta - t_c}\,(\eta - t)}\,d\eta\right) \tag{7}$$

where $X(t) = \sqrt{t - t_c}$ is a particular solution for the homogeneous solution in which the right hand side of the expression (6) is zero. We now choose a branch cut for X(t) such that

$$\begin{aligned}
&X^+(\eta) = \sqrt{\eta - t_c} && -\infty < \eta < t\\
&X^+(\eta) = -i\sqrt{t_c - \eta} && t_c < \eta < +\infty
\end{aligned} \tag{8}$$

In fact when the Froude numbers F_{r2} is infinite then the speed on the free surface CD′ is unity and therefore $\tau = 0$. Equation (7) then becomes

$$\Omega(t) = \frac{\sqrt{tc - t}}{\pi}\left(\int_{-\infty}^{-1} \frac{-\beta(\eta)}{\sqrt{tc - \eta}\,(\eta - t)}\,d\eta\right.$$

$$\left. - \frac{\pi}{2}\int_{-1}^{0} \frac{1}{\sqrt{tc - \eta}\,(\eta - t)}\,d\eta + \frac{\pi}{2}\int_{0}^{tc} \frac{1}{\sqrt{tc - \eta}\,(\eta - t)}\,d\eta\right) \qquad (9)$$

In equation (9) we let t be in the upper half plane and then tend to the real axis η and take the Cauchy Principal Value. Separating the real and imaginary parts of Ω, then the angle $\theta(t)$ on the free surface AD is given by:

$$\theta(t) = \frac{\sqrt{t - tc}}{\pi}\int_{-\infty}^{-1} \frac{-\beta(\eta)}{\sqrt{tc - \eta}\,(\eta - t)}\,d\eta$$

$$+ \operatorname{arctg}\sqrt{\frac{t_c + 1}{t - t_c}} - 2\operatorname{arctg}\sqrt{\frac{t_c}{t - t_c}} \qquad -\infty < t < -1 \qquad (10)$$

The fluid speed on AD is then given by

$$\operatorname{Ln} u(t) = \tau(t) = \frac{\sqrt{tc - t}}{\pi} ⨍_{-\infty}^{-1} \frac{-\beta(\eta)}{\sqrt{tc - \eta}\,(\eta - t)}\,d\eta$$

$$+ \operatorname{Ln}\frac{\sqrt{t_c - t} + \sqrt{t_c}}{\sqrt{t_c - t} - \sqrt{t_c}} + \frac{1}{2}\operatorname{Ln}\frac{\sqrt{t_c - t} - \sqrt{t_c + 1}}{\sqrt{t_c - t} + \sqrt{t_c + 1}}$$

$$-\infty < t < -1 \qquad (11)$$

where $⨍$ is the Cauchy Principal Value.

The fluid speed on AS and SC are then given by

$$\operatorname{Ln} u(t) = \tau(t) = \frac{\sqrt{tc - t}}{\pi}\int_{-\infty}^{-1} \frac{-\beta(\eta)}{\sqrt{tc - \eta}\,(\eta - t)}\,d\eta$$

$$+ \operatorname{Ln}\frac{\sqrt{t_c - t} + \sqrt{t_c}}{\sqrt{t_c - t} - \sqrt{t_c}} + \frac{1}{2}\operatorname{Ln}\frac{\sqrt{t_c - t} - \sqrt{t_c + 1}}{\sqrt{t_c - t} + \sqrt{t_c + 1}}$$

$$-1 < t < -t_c \qquad (12)$$

Now consider the problem in the physical plane where the unknown functions are the profiles of the free surfaces AD and CD'. These include the positions of the cusp and the stagnation points, the speed of the fluid on SA and SC and the potential function on the boundary. On the free surfaces AD and CD we choose the arc length coordinates s, and the points A and C are the origin points, respectively. When we take the arc length s to be the independent variable, the angle, the speed of the fluid and the potential function on the free surfaces are functions of s and from equation (5) we have

$$d\eta = \pi\eta(s)u(s)ds \tag{13}$$

and

$$\frac{d\phi(s)}{ds} = u(s) \tag{14}$$

$$\left.\begin{aligned}\frac{dx(s)}{ds} &= \cos\theta(s)\\ \frac{dy(s)}{ds} &= \sin\theta(s)\end{aligned}\right\}\text{ on the free surfaces} \tag{15}$$

In the physical plane we may rewrite expressions (10), (11) and (12) as follows:

$$\theta(s) = \sqrt{t(s) - t_c}\int_{\Gamma_1}\frac{\beta(l)\eta(l)u(l)}{\sqrt{t_c - \eta(l)}\,[\eta(l)-t(s)]}\,dl + \operatorname{arctg}\sqrt{\frac{t_c + 1}{t(s) - t_c}} - 2\operatorname{arctg}\sqrt{\frac{t_c}{t(s) - t_c}} \qquad s \in \Gamma_2 \tag{16}$$

$$\operatorname{Ln} u(s) = \tau(s) = \sqrt{t_c - t(s)}\int_{\Gamma_1}\frac{\beta(l)\eta(l)u(l)}{\sqrt{t_c - \eta(l)}\,[\eta(l) - t(s)]}\,dl + \operatorname{Ln}\frac{\sqrt{t_c - t(s)} + \sqrt{t_c}}{\sqrt{t_c - t(s)} - \sqrt{t_c}} + \frac{1}{2}\operatorname{Ln}\frac{\sqrt{t_c - t(s)} - \sqrt{t_c + 1}}{\sqrt{t_c - t(s)} + \sqrt{t_c + 1}} \qquad s \in \Gamma_1 \tag{17}$$

$$\operatorname{Ln} u(t) = \tau(t) = \sqrt{t_c - t}\int_{\Gamma_1}\frac{\beta(l)\eta(l)u(l)}{\sqrt{t_c - \eta(l)}\,[\eta(l) - t]}\,dl + \operatorname{Ln}\frac{\sqrt{t_c - t} + \sqrt{t_c}}{\sqrt{t_c - t} - \sqrt{t_c}} + \frac{1}{2}\operatorname{Ln}\frac{\sqrt{t_c - t} - \sqrt{t_c + 1}}{\sqrt{t_c - t} + \sqrt{t_c + 1}} \qquad -1 < t < t_c \tag{18}$$

in which $t_C = e^{\pi\phi(C)}$, $\phi(C)$ is the potential function at the point C, and on Γ_1, $\eta(l) = -e^{\pi\phi(l)}$ and $t(s) = -e^{\pi\phi(s)}$ whilst on Γ_2, $\eta(l) = e^{\pi\phi(l)}$, $t(s) = e^{\pi\phi(s)}$.

The potential functions on the free surfaces are given by

$$\phi(s) = \int_0^s u(s)ds \qquad \text{on } \Gamma_1 \tag{19}$$

$$\phi(s) = \phi_C + \int_0^s u(s)ds \qquad \text{on } \Gamma_2 \tag{20}$$

and the profile of the free surface CD′ is

$$\left.\begin{aligned} y(s) &= h_c + \int_0^s \sin\theta(s)ds \\ x(s) &= \int_0^s \cos\theta(s)ds \end{aligned}\right\} \quad \text{on } \Gamma_2 \qquad (22)$$

in which h_c is given, subject to the conditions that $y(s) \to 0$ as $s \to \infty$ and $h_A = \frac{F_{r1}^2}{2} + 1$. The directions of the linear integration along Γ_1 is along the direction of the velocity.

If the position C is known from the integration of equation (3) then the position of the source may be given by the expressions

$$h_s = h_c + \frac{1}{\pi}\int_0^{t_c} \frac{\sqrt{t_c} - \sqrt{t_c - t}}{t(\sqrt{t_c} + \sqrt{t_c + t})}\left(\frac{\sqrt{t_c + 1} - \sqrt{t_c - t}}{\sqrt{t_c + 1} - \sqrt{t_c - t}}\right)^{\frac{1}{2}} e^{-G(t)}dt$$

in which

$$G(t) = \sqrt{t_c - t}\int_{\Gamma} \frac{\beta(l)\eta(l)u(l)}{\sqrt{t - \eta(l)}\,[\eta(l) - t]}\,dl$$

NUMERICAL METHOD

Along Γ_1 the infinite integrals are truncated at some points M, say, (see Fig. 1), which are so far from the stagnation point that the angles, $\theta(s)$, at this point on the free surface is less than some preassigned small value and in all the examples taken in this paper it was found that 10^{-6} was sufficiently small. So the Cauchy Principal Value in equation (17) may be written as

$$\sqrt{t_c - t(s)} \oint_{\Gamma_1} \frac{\beta(l)\eta(l)u(l)}{\sqrt{t_c - \eta(l)}\,[\eta(l) - t(s)]}\,dl$$

$$= \sqrt{t_c - t(s)} \int_{\Gamma_1} \left(\frac{\beta(l)}{\sqrt{t_c - \eta(l)}} - \frac{\beta(s)}{\sqrt{t_c - t(s)}}\right)\frac{\eta(l)u(l)}{[\eta(l) - t(s)]}\,dl$$

$$+ \frac{\beta(s)}{\pi}\mathrm{Ln}\left(\frac{-1 - t(s)}{t(s) - t_M}\right) \qquad (36)$$

in which $t_M = -e^{\pi\phi(M)}$.

The mesh nodes on Γ_2 were distributed according to the magnitude of the change of angles in a similar manner to that in the finite element and boundary element methods. In the neighbourhood of the cusp point C the change is rapid and therefore the intervals are smaller than elsewhere. On the free surface AD, Γ_1, the lengths of the intervals are equal. The integration on each interval was evaluated using a Gaussian procedure.

In each computation the values of Fr1 and the potential function at the cusp point C were fixed. The aim of the numerical method is to obtain the function $\beta(s)$ on Γ_1, which is determined by equations (1) and (19). The solution $\beta(s)$ is identical to the function y(s), which is the profile of the free surface AD. A mesh of discrete s- and y-points is defined by $\{s_i\}$ and $\{y_i(s_i)\}$, and equations (1) and (19) give the nonlinear algebraic equations for the unknowns $\{y_i(s_i)\}$. A hybrid Powell method [18] can then be used to solve the nonlinear systems of equations. In all the calculations presented here the relative errors were less than 10^{-5} in the L_2 norm.

RESULTS

When the Froude numbers are Fr1 = 0.2 and Fr2 = ∞ then the solution will be only determined by the parameter ψ_c, which is the potential function at the cusp point C. We used the numerical procedure described in the above section to compute solutions for various values of the potential function ψ_c. The initial profiles of the free surface AC were taken to be horizontal planes. A solution was first obtained for a potential function ψ_c of -1, and solutions were then obtained for increasing the values of ψ_c, using the previous solution for the profile of the free surface as an initial guess for the next solution. The free surface profiles and the positions of source and cusp point for the values of ψ_c are shown in Fig. 3 and in table 1. As ψ_c increases then the position of the source and the cusp point C will move towards the stagnation point A. When the point S is far enough away from the stagnation point a small stagnation mount forms over the source. As the source gets closer to the stagnation point the free surface moves

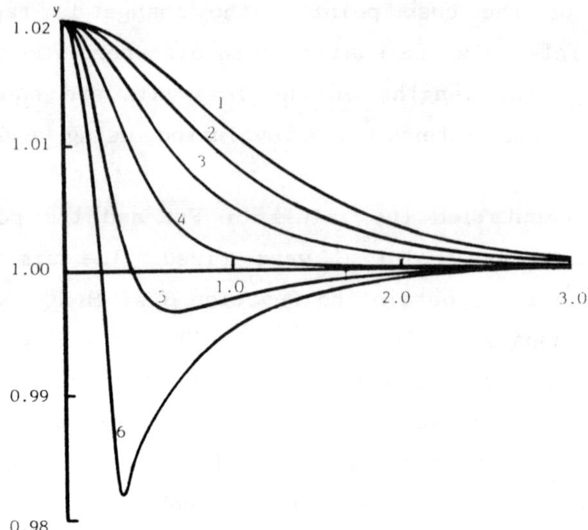

Fig. 3a The profiles of the free surfaces on which there is a stagnation point. 1— $\psi_c = -1$; 2— $\psi_c = -0.8$; 3— $\psi_c = -0.6$; 4— $\psi_c = -0.4$; 5— $\psi_c = -0.2$; 6— $\psi_c = 0.0$.

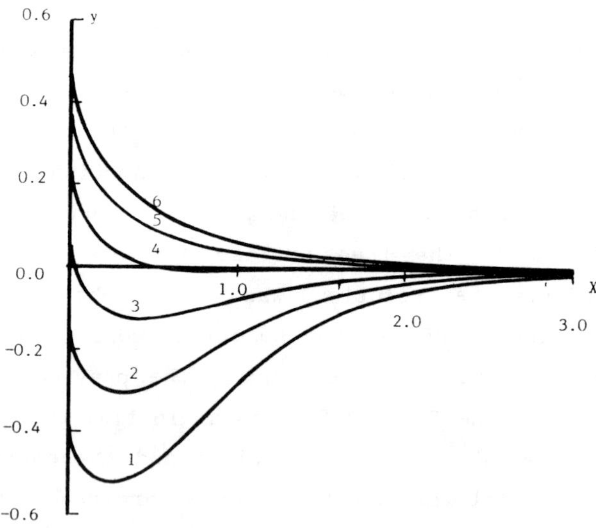

Fig. 3b The profiles of the free surfaces on which there is a cusp point. 1— $\psi_c = -1$; 2— $\psi_c = -0.8$; 3— $\psi_c = -0.6$; 4— $\psi_c = -0.4$; 5— $\psi_c = -0.2$; 6— $\psi_c = 0.0$.

towards the source and a trough of the free surface can be observed (see Fig. 3a) and its amplitude increases as the source approaches the stagnation point. We have found that solutions could be obtained for values of the ψ_c up to about a values of 0 where h_s=0.66898, h_c=0.46706 and when ψ_c>0 no solution could be found. The breakdown of the solution procedure physically indicates where the breakdown of the free surface should occur.

ψ_c	-1.0	-0.8	-0.6	-0.4	-0.2	0.0
h_s	-0.24803	-0.01001	0.20844	0.39837	0.55288	0.66898
h_c	-0.38706	-0.15532	0.05440	0.23235	0.37096	0.46706

Table 1 The positions of the source and cusp points as a function of the potential function at the cusp point.

The author is grateful to Professor Ingham for numerous discussions and invaluable suggestions.

REFERENCES

1. J. Imberger, J. C. Patterson, Physical Limnology, Advances in Applied Mechanics, Vol. 27, 302-475, 1989.
2. A. Craya, Theoretical research on the flow of nonhomogeneous fluids, La Houille Blanche **4**, 44-55 (1949).
3. C. S. Yih, Dynamics of Nonhomogeneous Fluids, Macmillan, New York, (1965).
4. D. H. Peregrine, A line source beneath a free surface, University of Wisconsin Report 1248, (1972).
5. J.-M. Vanden Broeck, L. W. Schwartz, and E. O. Tuck, Divergent low-Froude-number series expansion of non-linear free-surface flow problems, Proc. Roy. Soc. London Ser. **A 136**, 207-224 (1978).
6. G. C. Hocking and L. K. Forbes, A note on the flow induced by a line sink beneath a free surface, J . Austral. Math. Soc. Ser. B, 251-260 (1989).
7. H. Mekias and J.-M. Vanden-Broeck, Supercritical free-surface flow with a stagnation point due to a submerged source, Phys. Fluid A **1** (10), Oct. 1694-1697

(1989).

8. E. O. Tuck and J.-M. Vanden Broeck, A cusp-like free-surface flow due to a submerged source or sink, J. Austral. Math. Soc. Ser. B **25**, 443-450 (1984).
9. G. C. Hocking, Cusp-like free surface flows due to submerged source or sink in the presence of a flat or sloping bottom, J. Austral. Math. Soc. Ser. B **26**, 470-486, 1985.
10. I. L. Collings, Infinite Froude number cusped free surface flows due to a submerged line source or sink, J. Austral. Math. Soc. Ser. B **28**, 260-270 (1986).
11. J.-M. Vanden Broeck and J. B. Keller, Free surface flow due to a sink, J. Fluid Mech. **175**, 109-117 (1987).
12. G. C. Hocking, Infinite Froude number solutions to the problem of a submerged source or sink, J. Austral Math. Soc. Ser. B **29**, 401-409 (1988).
13. A. C. King and M. I. G. Bloor, A note on the free surface induced by a submerged source at infinite Froude number, J. Austral Math. Soc. Ser. B **30**, 147-156 (1988).
14. G. C. Hocking, Critical withdrawal from a two-layer fluid through a line sink, J. Eng. Maths. Vol. **25**, No. 1, 1-11, 1991.
15. X. Wen and D. B. Ingham, Flow induced by a submerged source or sink in a three-layer fluid, Computers and Fluids, in press.
16. X. Wen and D. B. Ingham, The free surface flow induced by a submerged source or sink from a three-layer fluid, Computational modelling of free and moving boundary problems, Vol. 1, 261-275, Edited by L. C. Wrobel and C. A. Brebbia, 1991.
17. N. I. Muskhelishvilli, Singular integral equations, Noordhoff, Groningen, The Netherlands, 1953.
18. M. J. D. Powell, A hybrid method for nonlinear algebraic equation, Numerical methods for nonlinear algebraic equations, P. Rabinowitz (ed), Gordon and Breach, 87-114, 1970.

Temperature Distribution on Flat Electronic Substrates in a Uniform Flow Field

S. De Smet, M. Driscart, G. De Mey
University of Ghent, Laboratory of Electronics, St.-Pietersnieuwstraat 41, B-9000 Ghent, Belgium

ABSTRACT

This paper deals with flat electronic substrates cooled by forced convection in a uniform flow field. Heat transport is performed by convection cooling to the ambient air and conduction in the substrate. The Boundary Element Method is used to solve this problem because this technique allows an elegant and efficient treatment of the governing equations.

INTRODUCTION

One of the keys to electronic system reliability is keeping integrated junction temperatures low. However, this simple design rule is easier said than done. Current trends in electronic design (increasing operating powers and speeds, ever growing scale integration, ...) make extracting heat and controlling temperatures more difficult. It is clear that thermal management is a crucial part of the system design. The distribution of the heat flux and the peak temperatures of power dissipating devices are of primary interest to the designer in determining the optimum placement of power electronic components on the substrate or the printed circuit board. The interest in thermal management is also reflected in the growing number of papers in this field (Mahalingam[2]) (Nakayama and Bergles[3]).

In this paper we consider a flat surface, such as a hybrid substrate or a printed circuit board, which is cooled by forced convection. Given a certain power distribution on the substrate, we are interested in the heat fluxes and substrate temperatures. As the substrates are rather thin (about $1mm$) and assuming that the components on the substrate are not too high, the air flow over the surface can be modelled as uniform and parallel to the substrate. It was

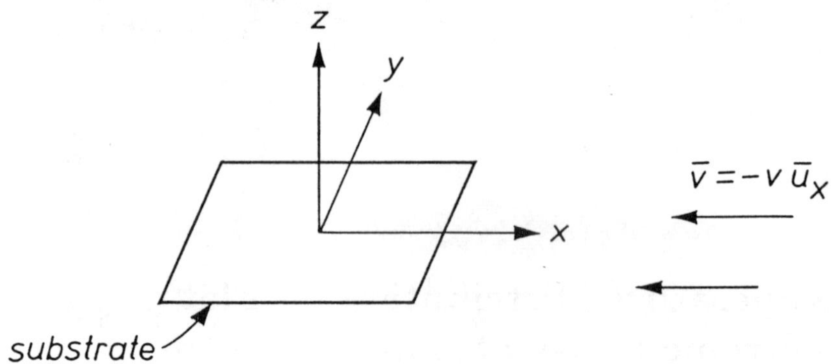

Figure 1: Ceramic substrate or printed circuit board in a uniform flow field

shown in a previous paper that the model of a uniform flow field yields results in good agreement with all well-known semi-analytical and empirical results in the literature (De Mey[1]).

The assumption of a known uniform flow field significantly simplifies the governing equations of this convection problem : the Navier-Stokes equations need not be solved. The remaining equations for the temperatures in the air and on the substrate can be solved elegantly and efficiently using the Boundary Element Method. Remark that in this paper a 3-D problem (diffusion-convection in the air) is coupled with a 2-D problem (conduction in the substrate).

INTEGRAL EQUATIONS

Consider a flat ceramic substrate or printed circuit board in the (x,y)-plane, in a uniform flow field $\overline{v}$ (fig.1) :

$$\overline{v} = -\, v\, \overline{u}_x \tag{1}$$

The cooling fluid can be air or an inert liquid. On the substrate, we have a given power distribution, caused by the heat dissipation of several electronic components. Two equations are governing this heat transfer problem :

- thermal conduction equation for the temperatures on the substrate
- thermal diffusion-convection equation for the fluid temperatures.

In a previous paper, the substrate was assumed at a uniform and known temperature (De Mey[1]). The thermal conduction equation was then disregarded.

Thermal diffusion-convection equation

The 3-D thermal diffusion-convection equation is given by :

$$\lambda_f \nabla^2 T_f = c_v \overline{v}.\nabla T_f \tag{2}$$

with

- T_f = fluid temperature in ^0C
- λ_f=thermal conductivity of the fluid in W/m^0C
- c_v=volumetric thermal capacity of the fluid in $J/m^{3\,0}C$

Taking into consideration the uniform flow field (1) we can rewrite (2) as :

$$\lambda_f \nabla^2 T_f = -\, c_v v \frac{\partial T_f}{\partial x} \tag{3}$$

For a detailed treatment of this equation the reader is referred to the literature (De Mey[1]), only the basic results are mentioned here.

The fundamental solution of (3) is given by Green's function $G_f(\overline{r} \mid \overline{r}')$:

$$G_f(\overline{r} \mid \overline{r}') = \frac{e^{(x'-x)/L}\, e^{-|\overline{r}-\overline{r}'|/L}}{4\pi \mid \overline{r} - \overline{r}' \mid} \tag{4}$$

with

- $\overline{r} = x\overline{u}_x + y\overline{u}_y + z\overline{u}_z$
- $L = \dfrac{2\lambda_f}{c_v v}$

L is called a characteristic length and is of order $10^{-6}m$ for air velocities of around $2m/s$.

Equation (3) can be transformed in an integral equation :

$$T_f(\overline{r}) = -2 \iint_S G_f(\overline{r} \mid \overline{r}') \frac{\partial T_f}{\partial z}(\overline{r}') dS' \tag{5}$$

where S denotes the upper surface of the substrate or printed circuit board.

Thermal conduction equation

On the substrate we have the 2-D thermal conduction equation :

$$\nabla^2 T_s = \frac{\partial^2 T_s}{\partial x^2} + \frac{\partial^2 T_s}{\partial y^2} = -\,2 \frac{\lambda_f}{\lambda_s t_s} \frac{\partial T_f}{\partial z} - \frac{p}{\lambda_s t_s} \tag{6}$$

with

- T_s = substrate temperature in ^0C

- λ_s=thermal conductivity of the substrate in W/m^0C
- t_s=thickness of the substrate or printed circuit board in m
- p=power density on the substrate in W/m^2

The fundamental solution of (6) - Green's function $G_s(\overline{r} \mid \overline{r}')$ - can be found by solving :

$$\nabla^2 G_s = -\delta(\overline{r} - \overline{r}') \tag{7}$$

yielding

$$G_s(\overline{r} \mid \overline{r}') = -\frac{1}{4\pi} ln(\mid \overline{r} - \overline{r}' \mid^2) \tag{8}$$

Applying Green's theorem to (6) and (7) gives :

$$\oint_{\partial S} (-T_s \frac{\partial G_s}{\partial n} + G_s \frac{\partial T_s}{\partial n}) dl = T_s - \frac{2\lambda_f}{\lambda_s t_s} \iint_S \frac{\partial T_f}{\partial z} G_s dS - \frac{1}{\lambda_s t_s} \iint_S p\, G_s dS \tag{9}$$

with

- S : upper surface of the substrate
- ∂S : boundary of the substrate.

On the boundary of the substrate (∂S) we have $\partial T_s / \partial n = 0$. To understand the physical meaning of this boundary condition, one has to realize that the substrate is so thin that almost no heat loss can occur along these sides. Substituting this boundary condition in (9) yields :

$$\begin{aligned} T_s(\overline{r}) \quad = \quad & \frac{2\lambda_f}{\lambda_s t_s} \iint_S \frac{\partial T_f}{\partial z}(\overline{r}') G_s(\overline{r} \mid \overline{r}') dS' \\ & + \frac{1}{\lambda_s t_s} \iint_S p(\overline{r}') G_s(\overline{r} \mid \overline{r}') dS' - \oint_{\partial S} T_s(\overline{r}') \frac{\partial G_s}{\partial n}(\overline{r} \mid \overline{r}') dl' \end{aligned} \tag{10}$$

On the substrate or printed circuit board (PCB) itself, the substrate temperature T_s should be equal to the fluid temperature T_f :

$$T_f = T_s$$

So, combining (5) and (10) finally gives :

$$\begin{aligned} \frac{-1}{\lambda_s t_s} \iint_S p(\overline{r}')\, G_s(\overline{r} \mid \overline{r}')\, dS' \quad = \quad & \iint_S \left[2\, G_f(\overline{r} \mid \overline{r}') + \frac{2\lambda_f}{\lambda_s t_s} G_s(\overline{r} \mid \overline{r}') \right] \frac{\partial T_f}{\partial z}(\overline{r}') dS' \\ & - \oint_{\partial S} T_s(\overline{r}') \frac{\partial G_s}{\partial n}(\overline{r} \mid \overline{r}') dl' \end{aligned} \tag{11}$$

It is the system of integral equations (10) and (11) that has to be solved numerically in order to obtain the substrate temperatures T_s on the boundary of the substrate ∂S and the temperature derivatives $\partial T_f / \partial z$ on the substrate

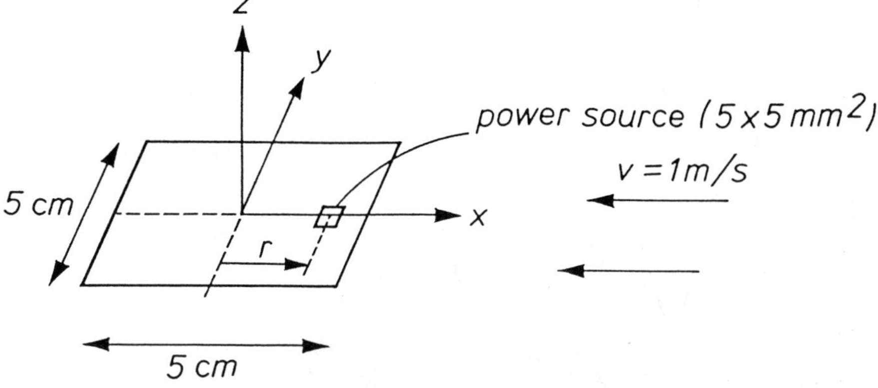

Figure 2: Ceramic substrate with power source

surface S. Once we have these values, we can calculate the temperature in any point of the substrate or fluid and the heat flux Q (in W/m^2) on the substrate which is defined by :

$$Q = -\lambda_f(\frac{\partial T_f}{\partial z})_{z=0} \tag{12}$$

For the numerical solution of the integral equations (10) and (11), the surface S has been divided in rectangular domains and the boundary ∂S in line segments. In each element the unknown T_s or $\frac{\partial T_f}{\partial z}$ is replaced by an unknown constant. The integrations involving G_f must be done very carefully as the characteristic length L is much smaller than the size of the surface elements (De Mey[1]). Further details about the numerical procedure are omitted here.

RESULTS

Hybrid circuit

Consider a square substrate with dimensions $5 \times 5cm^2$ and thickness $0.6mm$. These substrates are actually used for electronic hybrid circuits and they are made from ceramic materials such as Al_2O_3, AlN or BeO. Here we take Al_2O_3 with a thermal conductivity of $25W/m^0C$. Air is taken as the cooling medium with a velocity of $1m/s$. The thermal conductivity of air is $0.025W/m^0C$. On the substrate we consider a square shaped power source (modelling a heat dissipating electronic component) of $5 \times 5mm^2$, dissipating $1W$. The position of this source is varied along the x-axis, r being the x-coordinate of the centre of the source (fig.2).

Also note that we will talk about "temperature rises" instead of "temperatures" in this paragraph. The data on the graphs have to be interpreted as temperature rises relative to the ambient or free-stream air temperature.

On figure 3 the temperature rise on the substrate along the x-axis is plotted

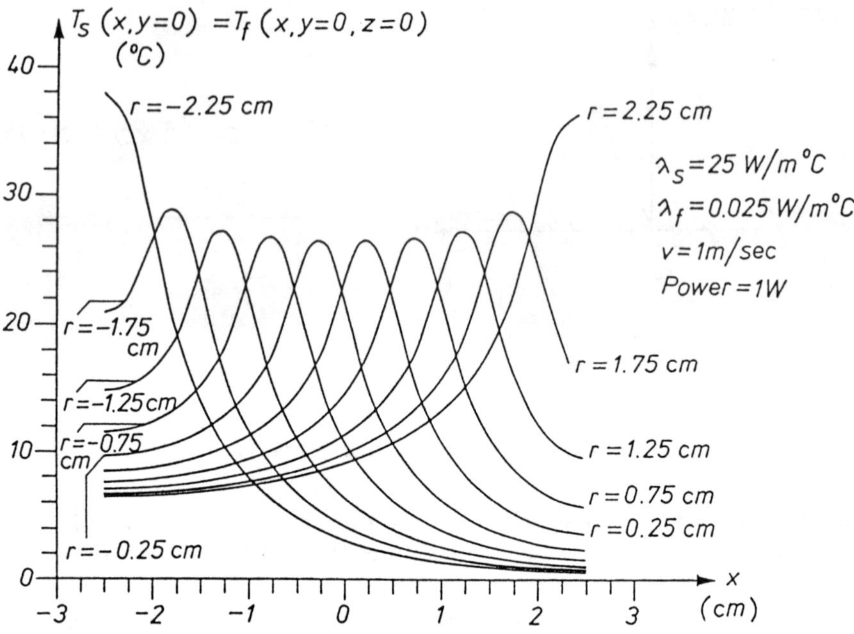

Figure 3: Temperature rise on ceramic substrate

for several positions r of the source. Two conclusions can be drawn from this graph. As a first conclusion, we can see that the peak temperatures are equal for symmetric positions r of the source. The peak temperature rise only depends on $|r|$ and not on r . So it doesn't matter wether the source is on the downstream or upstream side of the substrate. When the source is located at the high or low x-boundary of the substrate, we observe remarkable higher peak temperatures. This can be explained by a reduced heat conduction through the substrate in these cases. If we consider only one of these curves, we can clearly see the asymmetry in the temperature profile. This is due to the convection cooling. At the upstream side of the substrate, the air is still at ambient temperature, and we have a relatively small temperature rise on the substrate. As the air flows over the substrate, it becomes heated, causing higher temperature rises downstream.

This conclusion is also confirmed by fig.4 where the local heat flux $Q(x,0)$, as given by (12), along the x-axis is plotted for several thermal conductivities and for an almost central position of the source on the substrate. We can clearly see the higher heat flux on the upstream boundary, caused by the relatively cool air there. Downstream of the source we find a very low heat flux : the air has been heated and cannot remove as much heat as on the upstream side of the source.

A slight numerical instability can be seen at the upstream side of the heat flux profiles. These instabilities are also seen on the temperature profiles. They

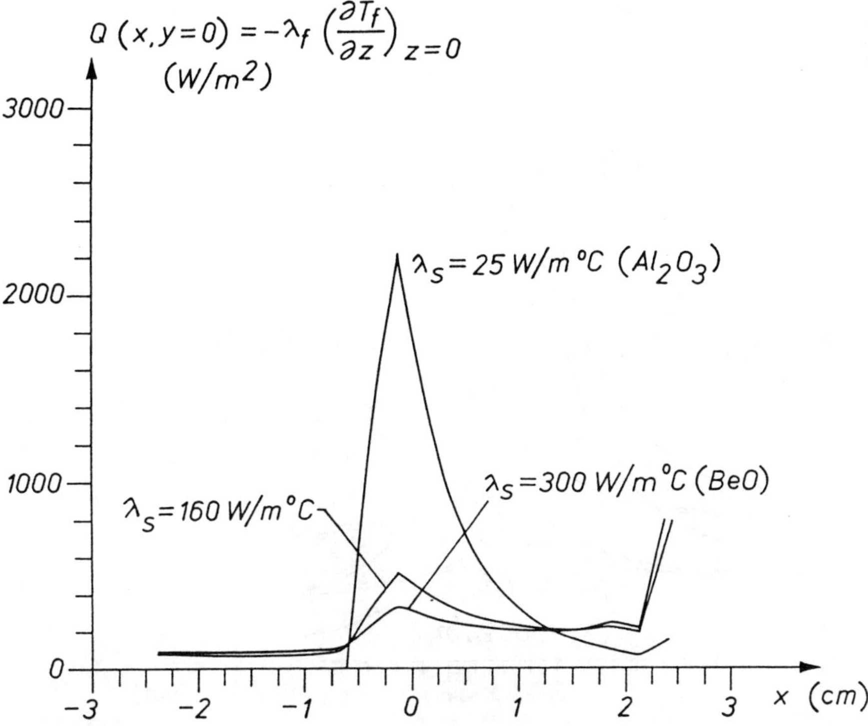

Figure 4: Heat flux

become smaller and even disappear for higher thermal conductivities. These instabilities are caused by the use of uniform boundary elements in which the unknown function is modelled as an (unknown) constant. The lower the thermal conductivity, the sharper the temperature profiles and the worse this model.

These simulations were made for different air velocities and some other ceramic materials (i.e. AlN with a thermal conductivity of $160W/m^0C$ and BeO with a thermal conductivity of $300W/m^0C$). Compared to the previous case, the curves giving the temperature rise and heat flux all have the same shape (fig.5). However some quantitative differences can be seen. A rather trivial observation is that higher air velocities result in smaller temperature rises. The higher the velocity, the better the cooling.
Substrates with a better thermal conductivity also give lower temperature rises, due to the better heat conduction. The better this thermal conductivity, the less sharp the temperature peak. The BeO-substrate with the highest thermal conductivity shows an almost isothermal temperature profile. High thermal conductivities give a better diffusion of the heat over the entire substrate and the temperature profiles become more smooth. This can also be seen on fig.4 where we have a smaller and less sharp heat flux peak for very good thermal conductors. In these cases, the total heat is conducted much better through the whole

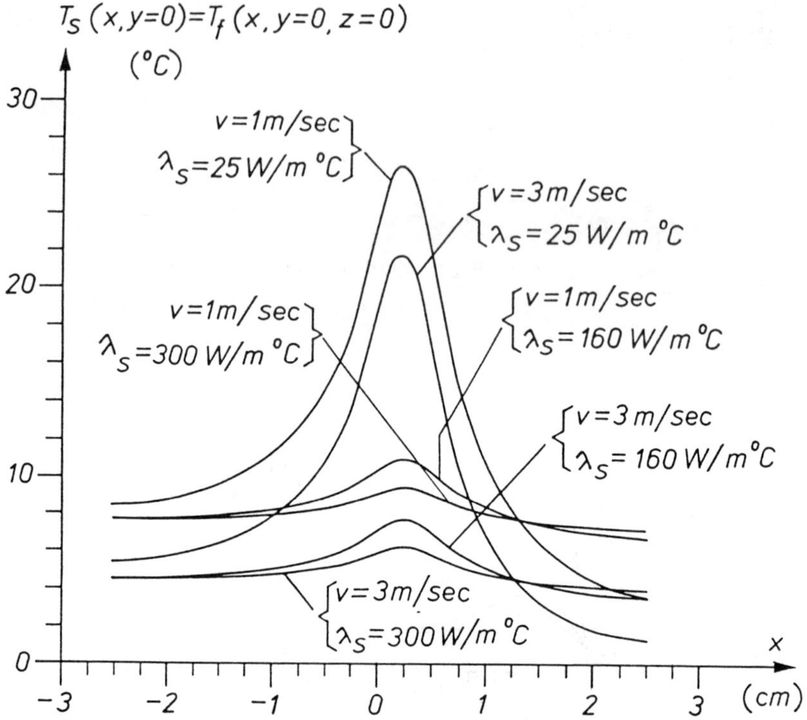

Figure 5: Comparison for different cases

substrate, the heat can be removed by the air from all over the substrate, resulting in a lower local heat flux on the source. For substrates with a low thermal conductivity, the heat is not spread as well, which results in a higher local heat flux in the source.

Printed circuit board

As a second example we will treat a printed circuit board (PCB) with dimensions $20 \times 20cm^2$ and thickness $1.6mm$. The thermal conductivities of the materials used for PCB's are considerably smaller than these used in hybrid circuit technology : ranging from $0.23W/m^0C$ for pure epoxy to $6.78W/m^0C$ when we take into account the copper conductors on the PCB (Wenthen[4]) (De Mey[5]). On the PCB we place a chip-carrier of $2cm$ by $2cm$ dissipating $1W$. If we vary the position of this power source along the x-axis, we observe almost the same phenomena as in the case of the ceramic substrates.

A completely different situation arises when this chip-carrier (dissipating $1W$) is superposed on a uniform power dissipation over the entire PCB (due to other heat dissipating components). Several power "backgrounds" are considered : $1W$, $5W$, $10W$ and $20W$ distributed uniformly over a PCB with a thermal conductivity of $6.78W/m^0C$. The air velocity is $3m/s$. For each of these cases, we first

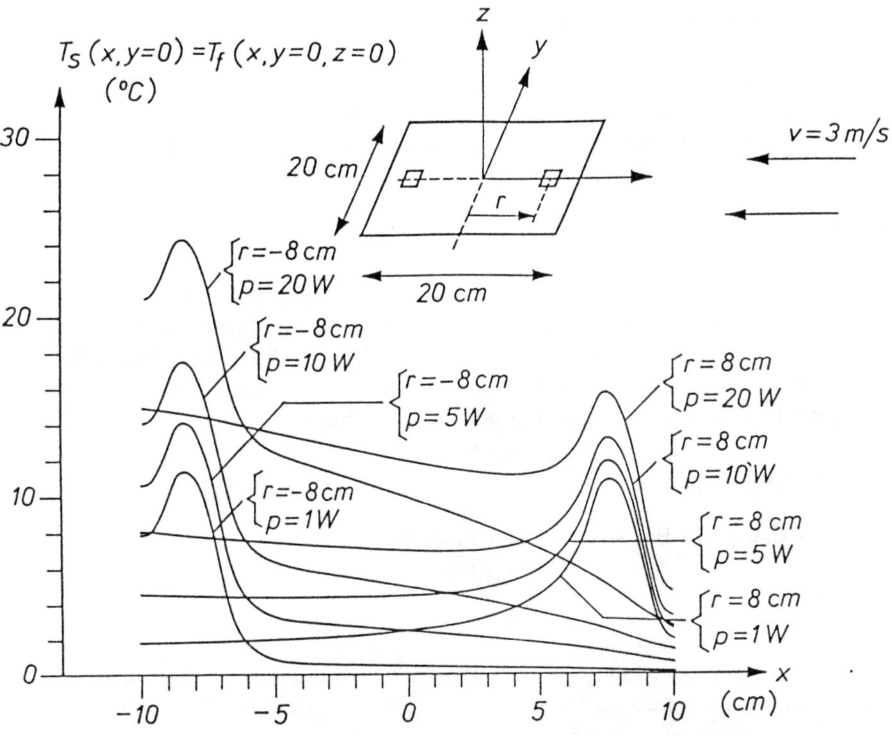

Figure 6: Temperature rise on PCB

	$r = -8cm$	$r = 8cm$
$1W$	11.45	10.99
$5W$	14.17	11.98
$10W$	17.56	13.26
$20W$	24.34	15.83

Table 1: Peak temperature rise in ^0C

place the chip-carrier on the upstream side ($r = 8cm$), and then on a symmetric position on the downstream side ($r = -8cm$) (fig.6). On figure 6 the temperature rise on the PCB along the x-axis is shown for these different situations. Here we do see a different peak temperature rise for symmetric positions of the chip-carrier. Table 1 lists the peak temperature rises for the various cases. Due to the uniform power dissipation over the whole PCB, the air is heated as it flows over the PCB. So, when the chip-carrier is on the downstream side, the air cannot remove as much heat as it could on the upstream side. When the chip-carrier is on the upstream side, it is cooled by air at ambient temperature. Hence the smaller peak temperature rise when the chip-carrier is located upstream. As the "background" power dissipation becomes higher, the difference in peak temperature rise for both positions also becomes more pronounced (table 1). This can be easily understood : the higher this dissipation, the more the air is preheated. These simulations show that components which dissipate a lot of heat should be placed at upstream positions of the PCB.

CONCLUSION

The conjugated heat transfer problem of cooling a flat substrate by forced convection, also taking into consideration the thermal conduction in it, has been solved successfully using the Boundary Element Method. This technique offers us a fast and accurate way to predict peak temperatures and heat fluxes on a substrate or printed circuit board in a uniform flow field.

ACKNOWLEDGMENTS

The authors thank the IWONL for their financial support.

REFERENCES

1. G. De Mey, S. De Smet, and M. Driscart. *Forced convection cooling of flat electronic substrates* in Betech/91 (Ed. : Brebbia C.A.), pp. 39-50, Proceedings of the Sixth Int. Conf. on Boundary Element Technology, Southampton, England, 1991. Computational Mechanics Publications, Southampton Boston, 1991.

2. M. Mahalingam. *Thermal management in semiconductor device packaging*, Proc. IEEE, Vol. 73, pp.1396-1404, 1985.

3. W. Nakayama and A.E. Bergles. *Cooling electronic equipment : past, present and future* in Heat Transfer in Electronic and Microelectronic Equipment (Ed. Bergles A.E.), pp. 3-39. Hemisphere Publishing Corporation, New York, 1990.

4. F. Wenthen. *The heat sink effect of printed conductors*, IEEE Trans. on Parts, Hybrids and Packaging, vol. PHP-12, no.2, pp.110-115, June 1976.

5. G. De Mey. *Thermal modelling of printed circuit boards vs. hybrid circuits*, Hybrid Circuit Technology, vol.4, no.1,pp.35-38, Jan.1987.

Computational Fluid Dynamics by the Boundary-Domain Integral Method

P. Skerget, A. Alujevic, I. Zagar, Z. Rek
Faculty of Engineering, University of Maribor, Slovenia

1 Introduction

Very fast development of computing enabled also the development of numerical fluid dynamics. It is numerical modelling and simulation of flow circumstances, including numerical experiments by the computer. Such procedure may have several important advantages over physical measurements on a laboratory model. It is of great importance that fluid properties (density, viscosity, compressibility, etc.) may be simply and arbitrarily changed, numerical experiment does not disturb the flow, plane flows can simply be simulated what may not be the case with laboratory experiments. The numerical experiment also has its own drawbacks and disadvantages, known to all numerical procedures, since the numerical solution represents a result of a discrete equation systems, which are not completely identical to basic physical laws of mechanics of continua. Discretisation often changes quantitatively and qualitatively the behavior of equations and thus also the solutions. Numerical simulation has also similar limitations like a laboratory experiments, since the solutions are individual discrete values only, not the functions of the flow fields.
Eventhough the numerical fluid dynamics has been recognised as an original attempt to the study of flow circumstances, it cannot totally replace physical experiments and theoretical analyses. Due to the difficulties of the subject, all these approaches are equally important and essential. Fluid dynamics is a research field, full of nonlinearities, strong geometrical nonregularities and singularities due to boundary conditions. Governing equations of transport phenomena are in general diffusivity-convectivity partial differential equations, the characteristics strongly change from point to point of the flow field, due to local Reynolds and Peclet number values, physically representing relationship between diffusion and convection of individual parameters of state. Thus it is not possible to discriminate pure elliptic, parabolic and hyperbolic equations, since they are of mixed

type. This particular character of equations makes the numerical fluid dynamics more difficult as compared to numerical solving of phenomena in solid.
Navier-Stokes equations represent a system of nonlinear partial differential equations of viscous Newtonian fluid motion. It is a mathematical model of physical conservation laws of mass, energy, species and momentum for a control volume – Eulerian case. In general it describes both laminar and turbulent flows. Since the turbulent flow is always space and time dependent, the direct simulation of turbulent flow by solving Navier-Stokes equations is practically impossible, engineering approach in most cases deals with simplified time averaged governing equations – Reynolds theory. Governing equations may be written for primitive physical variables or for dependent ones. For selection of the best formulation it is of great importance which numerical technique shall be applied. There is a variety of velocity-pressure, vorticity-stream function, velocity-vorticity, penalty formulations, etc. In particular the velocity-vorticity approach has shown its advantages with the boundary element method. Advantage of the velocity-vorticity formulation lies with the numerical separation of kinematics and kinetics of the flow from the pressure computation, which is determined later by the solution of a linear system of equations for known velocity and vorticity fields.
With most practical approximations only the steady state is of interest with respect to transient phenomena. It is common to all numerical techniques that they are usually more effective, also when determining the steady state, if this is achieved from time dependent solutions by a limit process. Time dependent system of equations is numerically simpler to deal with, it is stabler since in individual time steps has less nonlinear behavior. Transient case approach also does not presume an existance of steady state, which may always not even exist.

2 Integral Representation of Steady Diffusion-Convective Equation

2.1 Steady Diffusion-Convective Equation

The steady diffusion-convective equation represents an important class of partial differential equations, governing the steady transport phenomena in fluid flow, e.g. transfer of heat energy, momentum, vorticity, dispersion problems, etc. Due to the mixed elliptic-hyperbolic character of the mentioned PDE, the numerical solution of transport processes in fluids is much more difficult, than those in solids. This is specially true for flows characterised with high Reynolds or Peclet number values, when the convection becomes dominant compared with diffusion, or when the hyperbolic character of equation predominates the ellipticity of equation, respectively.
Let us consider a general steady state nonlinear diffusion-convective equation describing time nondependent transport of an arbitrary scalar function $u(\vec{r})$ in a homogenous, isotropic and incompressible medium of solution flow domain Ω bounded by the boundary Γ, e.g. given in indicial notation form for a right-handed Cartesian coordinate system

$$\rho_0 c \frac{\partial v_j u}{\partial x_j} = \frac{\partial}{\partial x_j}\left(\lambda \frac{\partial u}{\partial x_j}\right) + I_u \quad \text{in } \Omega , \tag{2.1.1}$$

where $\vec{v}(\vec{r})$ is the local solenoidal velocity field. The variable $u(\vec{r})$ can be interpreted, e.g. as a temperature in heat transfer problems, concentration in dispersion processes, vorticity in fluid dynamics problems etc., and will be refered to as a potential. The material properties of media are in general case functions of the potential and space, e.g. capacity $c = c(\vec{r}, u)$ and conductivity $\lambda = \lambda(\vec{r}, u)$, while $I_u(\vec{r}, u)$ stands for some source or sink term. In heat transfer problems for example c, λ, ρ_0 and I_u stand for specific isobaric heat, heat conductivity, constant fluid density and heat source respectively, while the thermal diffusivity is defined as $a = \lambda/\rho c$. The eq. (2.1.1) represents an elliptic boundary values problem, thus some boundary conditions have to be specified too, to complete the mathematical description of the transport problem, e.g. Dirichlet, Neumann or Cauchy type boundary conditions have to be known on the part of the boundary Γ_1, Γ_2 and Γ_3

$$\begin{aligned} u &= \overline{u} && \text{on } \Gamma_1 , \\ \frac{\partial u}{\partial x_j} n_j &= \overline{\frac{\partial u}{\partial n}} && \text{on } \Gamma_2 , \\ \frac{\partial u}{\partial x_j} n_j &= \alpha_u (u - u_f) && \text{on } \Gamma_3 , \end{aligned} \tag{2.1.2}$$

where α_u is transfer coefficient between the fluid flow surface defined by the unit normal vector $\vec{n}$, and the surrounding ambient at the potential u_f.

2.1.1 Linear Diffusion-Convective Equation

For the sake of simplicity, let us first consider an integral representation of a steady linear transport equation for a variable u in the known solenoidal velocity field $\vec{v}$

$$\frac{\partial v_j u}{\partial x_j} = a_0 \frac{\partial^2 u}{\partial x_j \partial x_j} + \frac{I_{u0}}{\rho_0 c_0} \quad \text{in } \Omega , \tag{2.1.3}$$

where a_0, ρ_0 and c_0 are constant material properties, while I_{u0} stands for known source term. The eq. (2.1.3) can be recognised to be a nonhomogeneous linear elliptic PDE of the form

$$\frac{\partial^2 u}{\partial x_j \partial x_j} + b_0 = 0 \quad \text{in } \Omega , \tag{2.1.4}$$

where b_0 represents the linear pseudo body force term. The corresponding boundary-domain integral representation to eq. (2.1.4) can be formulated by applying a weighted residual technique or simply by Green's theorems for scalar functions, resulting in the following integral statement

$$c(\xi)u(\xi) + \int_\Gamma u \frac{\partial u^{*E}}{\partial n} d\Gamma = \int_\Gamma \frac{\partial u}{\partial n} u^{*E} d\Gamma + \int_\Omega b_0 u^{*E} d\Omega . \tag{2.1.5}$$

The pseudo body force term b_0 stands for convective and source term from eq. (2.1.3)

$$b_0 = \frac{1}{a_0}\left(-\frac{\partial v_j u}{\partial x_j} + \frac{I_{u_0}}{\rho_0 c_0}\right) , \tag{2.1.6}$$

rendering equation

$$c(\xi)u(\xi) + \int_\Gamma u\frac{\partial u^{*E}}{\partial n}d\Gamma = \int_\Gamma \frac{\partial u}{\partial n}u^{*E}d\Gamma + \frac{1}{a_0}\int_\Omega \left(-\frac{\partial v_j u}{\partial x_j} + \frac{I_{u_0}}{\rho_0 c_0}\right) u^{*E} d\Omega , \tag{2.1.7}$$

where u^{*E} is the elliptic fundamental solution and $\partial u^*/\partial n$ its derivative in direction normal to the boundary, while $c(\xi)$ is the geometrically dependent free term accounting for the Cauchy type singularity of the integral on the left hand side of eq.(2.1.7).

The domain integral involves the partial derivatives of the field function u in convective term. They can be eliminated in various ways. One possibility is by performing the spatial derivation of eq. (2.1.7), what may not be economic due to the required solution of the additional integral equations at each interior point. The derivation of the shape functions, known in the finite elements formulation, is perhaps also not very appropriate, since this approach causes stability problems in the numerical solution scheme, and at the same time it is also not consistant in the frame of overall integral formulation. The third alternative is, which will be used herein, the use of the Gaussian divergence theorem on the convective domain integral part, e.g.

$$\int_\Omega \frac{\partial v_j u}{\partial x_j}u^{*E}d\Omega = \int_\Omega \frac{\partial v_j u u^{*E}}{\partial x_j}d\Omega - \int_\Omega u v_j \frac{\partial u^{*E}}{\partial x_j}d\Omega \tag{2.1.8}$$

and

$$\int_\Omega \frac{\partial v_j u}{\partial x_j}u^{*E}d\Omega = \int_\Gamma u v_n u^{*E}d\Gamma - \int_\Omega u v_j \frac{\partial u^{*E}}{\partial x_j}d\Omega , \tag{2.1.9}$$

resulting in an integral formulation without derivatives of field functions in the domain integral, where the divergence operator is shifted to the fundamental solution

$$c(\xi)u(\xi) + \int_\Gamma u\frac{\partial u^{*E}}{\partial n}d\Gamma = \int_\Gamma \frac{\partial u}{\partial n}u^{*E}d\Gamma - \frac{1}{a_0}\int_\Gamma u v_n u^{*E}d\Gamma$$

$$+ \frac{1}{a_0}\int_\Omega \left(u v_j \frac{\partial u^{*E}}{\partial x_j} + \frac{I_{u0}}{\rho_0 c_0}u^{*E}\right) d\Omega , \tag{2.1.10}$$

where $v_n = \vec{v}\cdot\vec{n}$ is the normal velocity component to the boundary. The eq. (2.1.10) is valid both for space ($j = 1, 2, 3$) as for plane ($j = 1, 2$) flow problems subject to use of appropriate space or plane elliptic fundamental solutions.

The boundary-domain integral representation eq. (2.1.10) describes the transport of the scalar function u in an integral form in a physically adequate manner. Due to this fact the numerical scheme resulting from the discretised integral equation is very stable and accurate. Notice, that the diffusion is a boundary problem only described by the first two boundary integrals, while the third boundary integral gives the resulting convective flux of the quantity u across the control surface, and which vanishes for zero normal velocity component, $v_n = 0$. The domain integral is due to the convective domain effects and the contribution of the source term on the development of the scalar field. For harmonic source this domain integral part can be transformed to an equivalent boundary integral.

2.1.2 Weak Nonlinear Diffusion-Convective Equation

In this text the term weak nonlinear indicates the nonlinearity of the governing equation which can be eliminated using simple Kirchhoff's integral transformation.

Let us consider the following nonlinear steady diffusion-convective equation

$$\rho_0 c \frac{\partial v_j u}{\partial x_j} = \frac{\partial}{\partial x_j}\left(\lambda \frac{\partial u}{\partial x_j}\right) + I_{u_0} \quad \text{in } \Omega \ , \tag{2.1.11}$$

where the nonlinearity exists basically due to the potential dependent conductivity $\lambda = \lambda(u)$. Applying the Kirchhoff's integral transformation defined by

$$\psi = \mathcal{K}[u] = \int_{u_0}^{u} \frac{\lambda(u)}{\lambda_0} du \tag{2.1.12}$$

and taking into the account the relation

$$\frac{\partial \psi}{\partial x_j} = \frac{\lambda}{\lambda_0} \frac{\partial u}{\partial x_j} \ , \tag{2.1.13}$$

the governing statement eq. (2.1.11) can be rewritten for the transformed variable $\psi(u)$

$$\frac{\partial v_j \psi}{\partial x_j} = a \frac{\partial^2 \psi}{\partial x_j \partial x_j} + \frac{a I_{u_0}}{\lambda_0} \quad \text{in } \Omega \ . \tag{2.1.14}$$

For the constant diffusivity, $a_0 = \lambda / \rho_0 c$, what is a reasonable good approximation of many engineering problems, the eq. (2.1.14) becomes linear

$$\frac{\partial v_j \psi}{\partial x_j} = a_0 \frac{\partial^2 \psi}{\partial x_j \partial x_j} + \frac{a_0 I_{u_0}}{\lambda_0} \quad \text{in } \Omega \ . \tag{2.1.15}$$

The transformed boundary conditions eq. (2.1.2) corresponding to eq. (2.1.15) are

$$\begin{aligned} \psi &= \overline{\psi} && \text{on} \quad \Gamma_1 \ , \\ \frac{\partial \psi}{\partial x_j} n_j &= \overline{\frac{\partial \psi}{\partial n}} && \text{on} \quad \Gamma_2 \ , \\ \frac{\partial \psi}{\partial x_j} n_j &= \frac{\alpha_u}{\lambda_0}\left(\mathcal{K}^{-1}[\psi] - u_f\right) && \text{on} \quad \Gamma_3 \ . \end{aligned} \tag{2.1.16}$$

Upon transformation the Dirichlet and Neumann boundary conditions remain linear, while the Cauchy mixed type boundary conditions become weak non-linear due to the inverse Kirchhoff's transform $\mathcal{K}^{-1}[\psi]$.
The eq. (2.1.15) can be again rewritten in a form of nonhomogeneous linear elliptic PDE

$$\frac{\partial^2 \psi}{\partial x_j \partial x_j} + b_0 = 0 \quad \text{in } \Omega \ , \tag{2.1.17}$$

where the linear pseudo body force term b_0 can be equated by

$$b_0 = \frac{1}{a_0}\left(-\frac{\partial v_j \psi}{\partial x_j} + \frac{I_{u_0}}{\lambda_0}\right) \ . \tag{2.1.18}$$

Due to the formal identity of eqs. (2.1.17) and (2.1.4) the corresponding integral statement can be easily formulated by replacing the variable u with ψ in eq. (2.1.10)

$$c(\xi)\psi(\xi) + \int_\Gamma \psi \frac{\partial u^{*\mathrm{E}}}{\partial n} d\Gamma = \int_\Gamma \frac{\partial \psi}{\partial n} u^{*\mathrm{E}} d\Gamma - \frac{1}{a_0}\int_\Gamma \psi v_n u^{*\mathrm{E}} d\Gamma$$

$$+ \frac{1}{a_0}\int_\Omega \left(\psi v_j \frac{\partial u^{*\mathrm{E}}}{\partial x_j} + \frac{I_{u0}}{\lambda_0} u^{*\mathrm{E}}\right) d\Omega \ . \tag{2.1.19}$$

By applying the Kirchhoff's transformation the weak nonlinearity in the governing diffusion-convective equation is effectively eliminated. For the Cauchy type boundary conditions the nonlinearity is shifted to the boundary conditions, what is still an appropriate mathematical description for the boundary-domain integral technique.

2.1.3 Nonlinear Diffusion-Convective Equation

In this case the nonlinearity in governing transport equation caused by non-constant material properties cannot be eliminated using Kirchhoff's integral transformation.
The integral representation of the following nonlinear diffusion-convective equation

$$\frac{\partial v_j u}{\partial x_j} = \frac{\partial}{\partial x_j}\left(a_e \frac{\partial u}{\partial x_j}\right) + \frac{I_u}{\rho_0 c_0} \quad \text{in } \Omega \ , \tag{2.1.20}$$

is discussed, where the effective diffusivity $a_e(\vec{r}, u)$ and the source term $I_u(\vec{r}, u)$ are some monotonic space and potential dependent functions. The effective diffusivity a_e can be always partitioned into a constant a_0 and a variable part $a_N(\vec{r}, u)$

$$a_e = a_0 + a_N(\vec{r}, u) \ . \tag{2.1.21}$$

By introducing expression (2.1.21) into eq. (2.1.20) the following statement can be written

$$\frac{\partial v_j u}{\partial x_j} = a_0 \frac{\partial^2 u}{\partial x_j \partial x_j} + \frac{\partial}{\partial x_j}\left(a_N \frac{\partial u}{\partial x_j}\right) + \frac{I_u}{\rho_0 c_0} \quad \text{in } \Omega \ , \tag{2.1.22}$$

which can be again recognised to be an elliptic nonhomogeneous PDE of the form

$$\frac{\partial^2 u}{\partial x_j \partial x_j} + b = 0 \quad \text{in } \Omega \ , \tag{2.1.23}$$

where the nonlinear pseudo body force term b includes the convection, nonlinear diffusion part and source term

$$b = \frac{1}{a_0}\left[-\frac{\partial}{\partial x_j}\left(v_j u - a_N \frac{\partial u}{\partial x_j}\right) + \frac{I_u}{\rho_0 c_0}\right] \ . \tag{2.1.24}$$

Due to the formal similarity between linear and nonlinear diffusion-convective equation, the integral formulation for the nonlinear one, may be directly rewritten as

$$c(\xi)u(\xi) + \int_\Omega u \frac{\partial u^{*\mathrm{E}}}{\partial n} d\Gamma = \int_\Gamma \frac{\partial u}{\partial n} u^{*\mathrm{E}} d\Gamma$$

$$+ \frac{1}{a_0}\int_\Omega \left[-\frac{\partial}{\partial x_j}\left(v_j u - a_N \frac{\partial u}{\partial x_j}\right) + \frac{I_u}{\rho_0 c_0}\right] u^{*\mathrm{E}} d\Omega \ . \tag{2.1.25}$$

The divergence operator in the domain integral can be shifted to the fundamental solution as before by applying Gaussian theorem, resulting in the following boundary-domain integral representation

$$c(\xi)u(\xi) + \int_\Omega u \frac{\partial u^{*\mathrm{E}}}{\partial n} d\Gamma = \int_\Gamma \frac{a_e}{a_0}\frac{\partial u}{\partial n} u^{*\mathrm{E}} d\Gamma - \frac{1}{a_0}\int_\Gamma u v_n u^{*\mathrm{E}} d\Gamma$$

$$+ \frac{1}{a_0}\int_\Omega \left[\left(u v_j - a_N \frac{\partial u}{\partial x_j}\right)\frac{\partial u^{*\mathrm{E}}}{\partial x_j} + \frac{I_u}{\rho_0 c_0} u^{*\mathrm{E}}\right] d\Omega \ . \tag{2.1.26}$$

2.2 Numerical Solution of Diffusion-Convection Equation

For the numerical solution of diffusion-convective equation the corresponding boundary-domain integral representation is written in a discretised form in which the integrals over the boundary and domain are approximated by a summation of integrals over individual boundary elements or internal cells respectively, e.g. in general discretised representation for the nonlinear integral statement eq. (2.1.26)

$$c(\xi)u(\xi) + \sum_{e=1}^{E}\int_{\Gamma_e} u \frac{\partial u^{*\mathrm{E}}}{\partial n} d\Gamma = \sum_{e=1}^{E}\int_{\Gamma_e} \frac{a_e}{a_0}\frac{\partial u}{\partial n} u^{*\mathrm{E}} d\Gamma - \frac{1}{a_0}\sum_{e=1}^{E}\int_{\Gamma_e} u v_n u^{*\mathrm{E}} d\Gamma$$

$$+ \frac{1}{a_0}\sum_{c=1}^{C}\int_{\Omega_c}\left[\left(u v_j - a_N \frac{\partial u}{\partial x_j}\right)\frac{\partial u^{*\mathrm{E}}}{\partial x_j} + \frac{I_u}{\rho_0 c_0} u^{*\mathrm{E}}\right] d\Omega \ , \tag{2.2.1}$$

where E boundary elements and C internal cells are employed. Clearly, for the constant diffusivity, $a = a_0$, $a_N = 0$, and for the linear source term $I_u = I_{u_0}$, the above equation reduces to a linear one. Next, the variation of all field functions or the products of field functions within each boundary element or internal cell is approximated by the use of interpolation polynomials $\{\Phi\}$ and $\{\phi\}$ with respect to boundary or domain and nodal function values. Let the index n refer to the number of nodes in each boundary element or internal cell, and also relate to the degree of the respective interpolation polynomials. It should be mentioned, that the degree of interpolation polynomials on the boundary elements and in internal cells can differ. With respect to the approximations employed different solution procedures can be developed, e.g. the implicit-explicit and implicit numerical scheme.

The implicit-explicit numerical procedure indicates, that the implicit system of equations is written for the boundary unknowns only, while all computations in the domain are performed on an explicit manner, point by point. This numerical solution approach has inherent nonlinearity due to the partition of the computation on implicit and explicit part, and the numerical solution even for original linear diffusion-convection case is achieved only through some iterative point under-relaxation procedure. Due to the fact that the system matrix is basicaly influenced only by diffusion, while the convection appears as a known system right hand side, the numerical scheme is less stable and generally diverges for flows with high Reynolds or Peclet number values.

The alternative implicit algorithm indicates that the implicit system of equations is written simultaneously for all boundary and internal points, resulting in a very large system matrix, which is now influenced by diffusion and convection. Consequence of this fact is that the numerical scheme is very stable and accurate regardless of Reynolds or Peclet number values.

For the solution of practical diffusion-convective problems, the sub-structure and clustering techniques have to be used due to enormous computer and memory demands.

2.2.1 Iterative Implicit-Explicit Scheme

The following approximations of functions and products of functions respectively, on the boundary and interior of domain are used

$$\begin{aligned} u &= \{\Phi\}^T\{u\}^n \ , \\ \frac{a_e}{a_0}\frac{\partial u}{\partial n} &= \{\Phi\}^T\left\{\frac{a_e}{a_0}\frac{\partial u}{\partial n}\right\}^n \ , \\ uv_n &= \{\Phi\}^T\{uv_n\}^n \ , \\ uv_j - a_N\frac{\partial u}{\partial x_j} &= \{\phi\}^T\left\{uv_j - a_N\frac{\partial u}{\partial x_j}\right\}^n \ , \\ \frac{I_u}{\rho_0 c_0} &= \{\phi\}^T\left\{\frac{I_u}{\rho_0 c_0}\right\}^n \ , \end{aligned} \qquad (2.2.2)$$

where T stands for the transposition. With the expressions (2.2.2) the discretised integral formulation eq.(2.2.1) may be written on the following manner

$$c(\xi)u(\xi)+\sum_{e=1}^{E}\left[\int_{\Gamma_e}\{\Phi\}^T\frac{\partial u^{*E}}{\partial n}d\Gamma\right]\{u\}^n=\sum_{e=1}^{E}\left[\int_{\Gamma_e}\{\Phi\}^T u^{*E}d\Gamma\right]\left\{\frac{a_e}{a_0}\frac{\partial u}{\partial n}\right\}^n$$

$$-\frac{1}{a_0}\sum_{e=1}^{E}\left[\int_{\Gamma_e}\{\Phi\}^T u^{*E}d\Gamma\right]\{uv_n\}^n+\frac{1}{a_0}\sum_{c=1}^{C}\left[\int_{\Omega_c}\{\phi\}^T\frac{\partial u^{*E}}{\partial x_j}d\Omega\right]\left\{uv_j-a_N\frac{\partial u}{\partial x_j}\right\}^n$$

$$+\frac{1}{a_0}\sum_{c=1}^{C}\left[\int_{\Omega_c}\{\phi\}^T u^{*E}d\Omega\right]\left\{\frac{I_u}{\rho_0 c_0}\right\}^n . \tag{2.2.3}$$

With the space integrals

$$h_e^n=\int_{\Gamma_e}\Phi^n\frac{\partial u^{*E}}{\partial n}d\Gamma \quad , \quad g_e^n=\int_{\Gamma_e}\Phi^n u^{*E}d\Gamma \ ,$$

$$d_{c_j}^n=\int_{\Omega_c}\phi^n\frac{\partial u^{*E}}{\partial x_j}d\Omega \quad , \quad d_c^n=\int_{\Omega_c}\phi^n u^{*E}d\Omega \ , \tag{2.2.4}$$

which are functions of the geometry only, the discretised eq. (2.2.3) is transformed into the following statement

$$c(\xi)u(\xi)+\sum_{e=1}^{E}\{h\}^T\{u\}^n=\sum_{e=1}^{E}\{g\}^T\left\{\frac{a_e}{a_0}\frac{\partial u}{\partial n}\right\}^n-\frac{1}{a_0}\sum_{e=1}^{E}\{g\}^T\{uv_n\}^n$$

$$+\frac{1}{a_0}\sum_{c=1}^{C}\{d_j\}^T\left\{uv_j-a_N\frac{\partial u}{\partial x_j}\right\}^n+\frac{1}{a_0}\sum_{c=1}^{C}\{d\}^T\left\{\frac{I_u}{\rho_0 c_0}\right\}^n . \tag{2.2.5}$$

The equation has to be first applied to all boundary nodes using standard collocation technique, $\xi = 1, N_e$ where N_e stands for the number of all boundary nodes, rendering an implicit system of equations for the boundary unknowns only, $\{u\}$ or $\{a_e/a_0\ \partial u/\partial n\}$ respectively, e.g. written in a matrix notation

$$[c(\xi)]\{u(\xi)\}+[\widehat{H}]\{u\}=[G]\left\{\frac{a_e}{a_0}\frac{\partial u}{\partial n}\right\}-\frac{1}{a_0}[G]\{uv_n\}$$

$$+\frac{1}{a_0}[D_j]\left\{uv_j-a_N\frac{\partial u}{\partial x_j}\right\}+\frac{1}{a_0}[D]\left\{\frac{I_u}{\rho_0 c_0}\right\} , \tag{2.2.6}$$

or

$$[H]\{u\}=[G]\left\{\frac{a_e}{a_0}\frac{\partial u}{\partial n}\right\}-\frac{1}{a_0}[G]\{uv_n\}$$

$$+\frac{1}{a_0}[D_j]\left\{uv_j-a_N\frac{\partial u}{\partial x_j}\right\}+\frac{1}{a_0}[D]\left\{\frac{I_u}{\rho_0 c_0}\right\} , \tag{2.2.7}$$

where the diagonal matrix of geometrical coefficients $[c(\xi)]$ is added to the matrix $[\widehat{H}]$. The term on the left hand side and the first term on the right hand side of eq. (2.2.7) describe diffusion, while the other terms represent boundary and domain convection effects, which are nonlinear due to the products of the field functions $\{uv_n\}$ and $\{uv_j\}$, nonlinear diffusion and nonlinear source. Thus, the following implicit nonlinear system of N_e equations for boundary unknowns can be formed

$$[H]\{u\}^{k+1} = [G]\left\{\frac{a_e}{a_o}\frac{\partial u}{\partial n}\right\}^{k+1} + \{F_N(\vec{v},u)\}^k \ , \tag{2.2.8}$$

where k stands for the iterative step.
Introducing the boundary conditions, e.g. prescribed nodal potential values $\{u\}$ on Γ_1 and nodal flux values $\{\frac{a_e}{a_0}\frac{\partial u}{\partial n}\}$ on Γ_2 part of the boundary $\Gamma = \Gamma_1 + \Gamma_2$, the eq. (2.2.8) can be reordered in such a way, that all the unknowns are on the left hand side, resulting in the following set of equations

$$[A]\{X\}^{k+1} = \{F_o\} + \{F_N(\vec{v},u)\}^k \ . \tag{2.2.9}$$

The system matrix $[A]$ is composed from the $[G]$ and $[H]$ influenced matrices, e.g.

$$[A] = [-[G], [H]] \ , \tag{2.2.10}$$

while the vector of unknowns $\{X\}$ is

$$\{X\} = \left\{ \begin{array}{c} \left\{\frac{a_e}{a_0}\frac{\partial u}{\partial n}\right\}^{\Gamma_1} \ , \\ \{u\}^{\Gamma_2} \end{array} \right\} \ . \tag{2.2.11}$$

The vector $\{F_o\}$ stands for the known boundary conditions, and the vector $\{F_N(\vec{v},u)\}$ expresses the contribution of convection and non-linear diffusion on the development of the potential field.
The next numerical step in solution procedure is an explicit computation of the potential values in the domain, $\xi = 1, N_c$, where N_c stands for the total number of all internal points. Taking $c(\xi) = 1$ in eq. (2.2.6) one can write

$$\begin{aligned} \{u(\xi)\}^{k+1} &= -[\widehat{H}]\{u\}^{k+1} + [G]\left\{\frac{a_e}{a_o}\frac{\partial u}{\partial n}\right\}^{k+1} - \frac{1}{a_o}[G]\{uv_n\}^k \\ &+ \frac{1}{a_o}[D_j]\left\{uv_j - a_N\frac{\partial u}{\partial x_j}\right\}^k + \frac{1}{a_o}[D]\left\{\frac{I_u}{\rho_o c_o}\right\}^k \ , \end{aligned} \tag{2.2.12}$$

or in the form

$$\{u(\xi)\}^{k+1} = -[\widehat{H}]\{u\}^{k+1} + [G]\left\{\frac{a_e}{a_o}\frac{\partial u}{\partial n}\right\}^{k+1} + \{F_N(\vec{v},u)\}^k \ . \tag{2.2.13}$$

For the large Peclet or Reynolds number values respectively, the nonlinearity incorporated in the vector $\{F_N(\vec{v},u)\}$ due to the product of velocity and potential, which exists even for the linear diffusion-convective equation, becomes more and more severe, making the numerical computation difficult. The under-relaxation point by point iterative method with reversed sweep of potential computation has to be taken into account for the explicit evaluation of the new domain potential values

$$u^{k+1} = Ku^{k+1} + (1-K)u^k \ . \tag{2.2.14}$$

The nonlinearity can be accounted for by decreasing the under-relaxation parameter K, $0 < K \leq 0.5$, or by a succesive increasing of the Peclet or Reynolds number values, which can considerably increases the stability of the numerical scheme. With the reversed sweep the convergency is accelerated and greater stability of the iterative algorithm is achieved. All new domain potential values are immediately considered in the nonlinear vector $\{F_N\}$. The iterative procedure is continued until some convergence criterion is satisfied, e.g.

$$\frac{\max_j |u^{k+1} - u^k|}{\max_j |u^{k+1}|} \leq \epsilon \quad \text{for } j = 1, N_c \ , \tag{2.2.15}$$

where ϵ is a small positive number, e.g. $\epsilon = 10^{-4}$ to 10^{-6}, depending on the desired accuracy.
As it is already pointed out, the implicit-explicit iterative procedure becomes unstable and the numerical solution diverges for higher Peclet or Reynolds number values. The reason lies in the fact, that in the system matrix $[A]$, only the diffusion transfer from the boundary is included, while the convection is treated as a nonlinear right hand side term, and since it has the dominant influence on the solution at higher Peclet or Reynolds number values, the numerical algorithm becomes badly conditioned.

2.2.2 Implicit Scheme

For this scheme the field functions in each boundary element and internal cell are to be approximated using the following expressions

$$\begin{aligned}
u &= \{\Phi\}^T\{u\}^n \ , \\
\frac{a_e}{a_o}\frac{\partial u}{\partial n} &= \{\Phi\}^T\left\{\frac{a_e}{a_o}\frac{\partial u}{\partial n}\right\}^n \ , \\
v_n &= \{\Phi\}^T\{v_n\}^n \ , \\
u &= \{\phi\}^T\{u\}^n \ , \\
v_j &= \{\phi\}^T\{v_j\}^n \ , \\
a_N\frac{\partial u}{\partial x_j} &= \{\phi\}^T\left\{a_N\frac{\partial u}{\partial x_j}\right\}^n \ , \\
\frac{I_u}{\rho_o c_o} &= \{\phi\}^T\left\{\frac{I_u}{\rho_o c_o}\right\}^n \ ,
\end{aligned} \tag{2.2.16}$$

yielding from eq. (2.2.1) the following discretised formulation

$$c(\xi)u(\xi) + \sum_{e=1}^{E}\left[\int_{\Gamma_e}\{\Phi\}^T\frac{\partial u^{*E}}{\partial n}d\Gamma\right]\{u\}^n = \sum_{e=1}^{E}\left[\int_{\Gamma_e}\{\Phi\}^T u^{*E}d\Gamma\right]\left\{\frac{a_e}{a_o}\frac{\partial u}{\partial n}\right\}^n$$

$$-\frac{1}{a_0}\sum_{e=1}^{E}\left[\int_{\Gamma_e}\{\Phi\}^T\left(\{\Phi\}^T\{v_n\}\right)u^{*\mathrm{E}}d\Gamma\right]\{u\}^n$$

$$+\frac{1}{a_0}\sum_{c=1}^{C}\left[\int_{\Omega_c}\{\phi\}^T\left(\{\phi\}^T\{v_j\}^n\frac{\partial u^{*\mathrm{E}}}{\partial x_j}\right)d\Omega\right]\{u\}^n$$

$$-\frac{1}{a_0}\sum_{c=1}^{C}\left[\int_{\Omega_c}\{\phi\}^T\frac{\partial u^{*\mathrm{E}}}{\partial x_j}d\Omega\right]\left\{a_N\frac{\partial u}{\partial x_j}\right\}^n+\frac{1}{a_0}\sum_{c=1}^{C}\left[\int_{\Omega_c}\{\phi\}^T u^{*\mathrm{E}}d\Omega\right]\left\{\frac{I_u}{\rho_0 c_0}\right\}^n . \tag{2.2.17}$$

The spatial integration terms become now in boundary elements and internal cells

$$\begin{aligned}
h_e^n &= \int_{\Gamma_e}\Phi^n\frac{\partial u^{*\mathrm{E}}}{\partial n}d\Gamma \quad , & g_e^n &= \int_{\Gamma_e}\Phi^n u^{*\mathrm{E}}d\Gamma \ , \\
c_e^n &= \int_{\Gamma_e}\Phi^n\left(\{\Phi\}^T\{v_n\}^n\right)u^{*\mathrm{E}}d\Gamma \quad , & \widehat{d}_c^n &= \int_{\Gamma_e}\phi^n\left(\{\phi\}^T\{v_j\}^n\right)\frac{\partial u^{*\mathrm{E}}}{\partial x_j}d\Omega \ , \\
d_{c_j}^n &= \int_{\Omega_c}\phi^n\frac{\partial u^{*\mathrm{E}}}{\partial x_j}d\Omega \quad , & d_c^n &= \int_{\Omega_c}\phi^n u^{*\mathrm{E}}d\Omega \ ,
\end{aligned} \tag{2.2.18}$$

Integrals eq. (2.2.18) are dependent upon geometry and velocity field. The discretised form of integral statement (2.2.17) can thus be written

$$c(\xi)u(\xi)+\sum_{e=1}^{E}\{h\}^T\{u\}^n=\sum_{e=1}^{E}\{g\}^T\left\{\frac{a_e}{a_0}\frac{\partial u}{\partial n}\right\}^n-\frac{1}{a_0}\sum_{e=1}^{E}\{c\}^T\{u\}^n$$

$$+\frac{1}{a_0}\sum_{c=1}^{C}\{\widehat{d}\}^T\{u\}^n-\frac{1}{a_0}\sum_{c=1}^{C}\{d_j\}^T\left\{a_N\frac{\partial u}{\partial x_j}\right\}^n+\frac{1}{a_0}\sum_{c=1}^{C}\{d\}^T\left\{\frac{I_u}{\rho_0 c_0}\right\}^n . \tag{2.2.19}$$

Applying eq. (2.2.19) to all bpundary and internal nodes $\xi = 1, N_e + N_c$, the following matrix equation is obtained

$$[c(\xi)]\{u(\xi)\}+[\widehat{H}]\{u\}=[G]\left\{\frac{a_e}{a_0}\frac{\partial u}{\partial n}\right\}-\frac{1}{a_0}[C]\{u\}$$

$$+\frac{1}{a_0}[\widehat{D}]\{u\}-\frac{1}{a_0}[D_j]\left\{a_N\frac{\partial u}{\partial x_j}\right\}+\frac{1}{a_0}[D]\left\{\frac{I_u}{\rho_0 c_0}\right\} , \tag{2.2.20}$$

or

$$[H]\{u\}=[G]\left\{\frac{a_e}{a_0}\frac{\partial u}{\partial n}\right\}-\frac{1}{a_0}[C]\{u\}$$

$$+\frac{1}{a_0}[\widehat{D}]\{u\}-\frac{1}{a_0}[D_j]\left\{a_N\frac{\partial u}{\partial x_j}\right\}+\frac{1}{a_0}[D]\left\{\frac{I_u}{\rho_0 c_0}\right\} . \tag{2.2.21}$$

Adding all corresponding terms together, one can write

$$\left([H] + \frac{1}{a_0}[C] - \frac{1}{a_0}[\widehat{D}]\right)\{u\} = [G]\left\{\frac{a_e}{a_0}\frac{\partial u}{\partial n}\right\} + \left\{F_N\left(u, \frac{\partial u}{\partial x_j}\right)\right\} \quad , \qquad (2.2.22)$$

or in a symbolic matrix notation of an implicit nonlinear system of $N_e + N_c$ equations for all boundary and domain unknown values

$$[E]\{u\}^{k+1} = [G]\left\{\frac{a_e}{a_0}\frac{\partial u}{\partial n}\right\}^{k+1} + \left\{F_N\left(u, \frac{\partial u}{\partial x_j}\right)\right\}^k \quad , \qquad (2.2.23)$$

where k stands for the iterative step. Notice, that for the constant material properties and linear source term, the above system of nonlinear equations reduces to linear one.

3 Integral Representation of Unsteady Diffusion-Convective Equation

3.1 Unsteady Diffusion-Convective Equation

The unsteady diffusion-convective equation represents mixed parabolic-hyperbolic type of partial differential equations, governing the time dependent transport phenomena in fluid flow. The time dependent transport statement for momentum, vorticity, temperature, concentration etc., can be recognised to be formally of the same type as a unsteady diffusion-convection equation.
Let us consider a general unsteady state nonlinear diffusion-convection equation describing time dependent transfer of an arbitrary scalar function $u(\vec{r},t)$ in a homogeneous, isotropic and incompressible fluid flow defined in solution domain $R = \Omega \times I$ representing the product of space Ω and time internal $I(t_o, t)$

$$\rho_o c\left(\frac{\partial u}{\partial t} + \frac{\partial v_j u}{\partial x_j}\right) = \frac{\partial}{\partial x_j}\left(\lambda\frac{\partial u}{\partial x_j}\right) + I_u \quad \text{in} \quad R\,. \qquad (3.1.1)$$

The variable $u(\vec{r},t)$ can be equated with temperature, concentration, vorticity etc., and will be refered to as a potential. The fluid material properties are in general potential and space dependent functions. The eq.(3.1.1) represents a parabolic initial-boundary values problem, thus some boundary and initial conditions have to be known to complete the mathematical description of the problem, e.g. Dirichlet, Neumann or Cauchy type boundary conditions have to be prescribed on the part of the boundary Γ_1, Γ_2 and Γ_3 respectively

$$\begin{aligned} u &= \overline{u} \quad &&\text{on} \quad \Gamma_1 \quad &&\text{for} \quad t > t_0\,, \\ \frac{\partial u}{\partial x_j}n_j &= \overline{\frac{\partial u}{\partial n}} \quad &&\text{on} \quad \Gamma_2 \quad &&\text{for} \quad t > t_0\,, \\ \frac{\partial u}{\partial x_j}n_j &= \alpha_u(u - u_f) \quad &&\text{on} \quad \Gamma_3 \quad &&\text{for} \quad t > t_0\,, \end{aligned} \qquad (3.1.2)$$

while the initial conditions are

$$u = \overline{u}_o \quad \text{in} \quad \Omega \quad \text{at} \quad t = t_o \,. \tag{3.1.3}$$

3.1.1 Linear Diffusion-Convective Equation

Let us first consider an integeral representation of a time dependent linear transport equation for an arbitrary scalar variable u in the known solenoidal velocity field $div\,\vec{v} = 0$

$$\frac{\partial u}{\partial t} + \frac{\partial v_j u}{\partial x_j} = a_o \frac{\partial^2 u}{\partial x_j \partial x_j} + \frac{I_{uo}}{\rho_o c_o} \quad \text{in} \quad R\,, \tag{3.1.4}$$

where a_o, ρ_o, c_o and I_{uo} are constant material properties and known source term respectively. The eq.(3.1.4) can be recognised to be a nonhomogeneous linear parabolic PDE of the form

$$a_o \frac{\partial^2 u}{\partial x_j \partial x_j} - \frac{\partial u}{\partial t} + b_o = 0 \quad \text{in} \quad R\,, \tag{3.1.5}$$

where b_o stands for linear pseudo body force term. The eq.(3.1.5) can be transformed into an equivalent boundary-domain integral equation by applying a weighted residual technique or Green's theorems for the scalar functions using the parabolic fundamental solution u^{*P}, e.g. written in the time incremental form for the time step $\tau = t_F - t_{F-1}$

$$c(\xi)u(\xi, t_F) + a_o \int_\Gamma \int_{t_{F-1}}^{t_F} u \frac{\partial u^{*P}}{\partial n} dt\, d\Gamma = a_o \int_\Gamma \int_{t_{F-1}}^{t_F} \frac{\partial u}{\partial n} u^{*P} dt\, d\Gamma$$

$$+ \int_{t_{F-1}}^{t_F} \int_\Omega b_o u^{*P} dt\, d\Omega + \int_\Omega u_{F-1} u^{*P}_{F-1} d\Omega \,. \tag{3.1.6}$$

The pseudo body force term b_o includes convection and source term from eq.(3.1.4)

$$b_o = -\frac{\partial v_j u}{\partial x_j} + \frac{I_{uo}}{\rho_o c_o}\,, \tag{3.1.7}$$

rendering an integral formulation

$$c(\xi)u(\xi, t_F) + a_o \int_\Gamma \int_{t_{F-1}}^{t_F} u \frac{\partial u^{*P}}{\partial n} dt\, d\Gamma = a_0 \int_\Gamma \int_{t_{F-1}}^{t_F} \frac{\partial u}{\partial n} u^{*P} dt\, d\Gamma$$

$$+ \int_\Omega \int_{t_{F-1}}^{t_F} \left(-\frac{\partial v_j u}{\partial x_j} + \frac{I_{uo}}{\rho_o c_o} \right) u^{*P} dt\, d\Omega + \int_\Omega u_{F-1} u^{*P}_{F-1} d\Omega \,. \tag{3.1.8}$$

Notice, that the direction derivatives of field functions exist in the first domain integral of eq.(3.1.8). This can be solved in various manners, e.g. by performing the spatial derivation of eq.(3.1.8), by derivation of shape functions, and by using Gauss divergence theorem on the convective domain integral part. This the last approach will be used herein, where the differential operator is shifted to the fundamental solution, e.g.

$$\int_\Omega \int_{t_{F-1}}^{t_F} \frac{\partial v_j u}{\partial x_j} u^{*P} dt\, d\Omega = \int_\Omega \int_{t_{F-1}}^{t_F} \frac{\partial v_j u u^{*P}}{\partial x_j} dt\, d\Omega - \int_\Omega \int_{t_{F-1}}^{t_F} u v_j \frac{\partial u^{*P}}{\partial x_j} dt\, d\Omega \tag{3.1.9}$$

and

$$\int_\Omega \int_{t_{F-1}}^{t_F} \frac{\partial v_j u}{\partial x_j} u^{*P} dt\, d\Omega = \int_\Gamma \int_{t_{F-1}}^{t_F} u v_n u^{*P} dt\, d\Gamma - \int_\Omega \int_{t_{F-1}}^{t_F} u v_j \frac{\partial u^{*P}}{\partial x_j} dt\, d\Omega\,, \tag{3.1.10}$$

resulting in the integral statement without derivatives of field functions within the domain integral

$$\begin{aligned} c(\xi)u(\xi,t_F) + a_o \int_\Gamma \int_{t_{F-1}}^{t_F} u \frac{\partial u^{*P}}{\partial n} dt\, d\Gamma &= a_o \int_\Gamma \int_{t_{F-1}}^{t_F} \frac{\partial u}{\partial n} u^{*P} dt\, d\Gamma \\ - \int_\Gamma \int_{t_{F-1}}^{t_F} u v_n u^{*P} dt\, d\Gamma &+ \int_\Omega \int_{t_{F-1}}^{t_F} \left(u v_j \frac{\partial u^{*P}}{\partial x_j} + \frac{I_{uo}}{\rho_o c_o} \right) dt\, d\Omega \\ &+ \int_\Omega u_{F-1} u^{*P}_{F-1}\, d\Omega\,. \end{aligned} \tag{3.1.11}$$

The eq.(3.1.11) is adequate both for space ($j = 1, 2, 3$) as for plane ($j = 1, 2$) flow geometry subject to use of appropriate space or plane parabolic fundamental solutions.

The boundary-domain integral statement eq.(5.1.11) represents time dependent transport of scalar function u in the integral form on a physicaly justified way. The diffusion process is described by the first two boundary integrals, while the third boundary integral represents the convective flow on the boundary which vanishes for zero normal velocity component $\dot{v}_n = 0$. The first domain integral is due to the convection and source, while the last domain integral gives the initial conditions effects on the development of the potential field in the next time intervale.

3.1.2 Nonlinear Diffusion-Convective Equation

In this case the nonlinearity in unsteady transfer equation caused by nonlinear physical properties can not be solved by Kirchhoff's integral transformation.
The integral representation of the following unsteady nonlinear diffusion-convective equation

$$\frac{\partial u}{\partial t} + \frac{\partial v_j u}{\partial x_j} = \frac{\partial}{\partial x_j}\left(a_e \frac{\partial u}{\partial x_j} \right) + \frac{I_u}{\rho_o c_o} \quad \text{in} \quad R \tag{3.1.12}$$

is developed, where the effective diffusivity $a_e(\vec{r}, u)$ and source term $I_u(\vec{r}, u)$ are same arbitrary space and potential dependent functions. Substituting the expression for effective diffusivity variation of a form

$$a_e = a_o + a_{\rm N}(\vec{r}, u) \,, \tag{3.1.13}$$

eq.(3.1.12) can be partitioned into a linear and nonlinear part in the following manner

$$\frac{\partial u}{\partial t} + \frac{\partial v_j u}{\partial x_j} = a_o \frac{\partial^2 u}{\partial x_j \partial x_j} + \frac{\partial}{\partial x_j}\left(a_{\rm N}\frac{\partial u}{\partial x_j}\right) + \frac{I_u}{\rho_o c_o} \quad \text{in} \quad R \,, \tag{3.1.14}$$

which can be seen as a nonhomogeneous parabolic PDE of the form

$$a_o \frac{\partial^2 u}{\partial x_j \partial x_j} - \frac{\partial u}{\partial t} + b = 0 \quad \text{in} \quad R \,. \tag{3.1.15}$$

Equating the pseudo body force term b with convection, nonlinear diffusion and source term

$$b = -\frac{\partial}{\partial x_j}\left(v_j u - a_{\rm N}\frac{\partial u}{\partial x_j}\right) + \frac{I_u}{\rho_o c_o} \,, \tag{3.1.16}$$

one can derive the following integral statement

$$c(\xi)u(\xi, t_F) + a_o \int_\Gamma \int_{t_{F-1}}^{t_F} u \frac{\partial u^{*\rm P}}{\partial n} dt\, d\Gamma = a_o \int_\Gamma \int_{t_{F-1}}^{t_F} \frac{\partial u}{\partial n} u^{*\rm P} dt\, d\Gamma$$

$$+ \int_\Omega \int_{t_{F-1}}^{t_F} \left[-\frac{\partial}{\partial x_j}\left(v_j u - a_{\rm N}\frac{\partial u}{\partial x_j}\right) + \frac{I_u}{\rho_o c_o}\right] u^{*\rm P} dt\, d\Omega + \int_\Omega u_{F-1} u^{*\rm P}_{F-1} d\Omega \,. \tag{3.1.17}$$

The differential operator in the first domain integral can be shifted to the parabolic fundamental solution by using Gaussian theorem, resulting in the following-boundary domain integral representation

$$c(\xi)u(\xi, t_F) + a_o \int_\Gamma \int_{t_{F-1}}^{t_F} u \frac{\partial u^{*\rm P}}{\partial n} dt\, d\Gamma = \int_\Gamma \int_{t_{F-1}}^{t_F} a_e \frac{\partial u}{\partial n} u^{*\rm P} dt\, d\Gamma$$

$$- \int_\Gamma \int_{t_{F-1}}^{t_F} u v_n u^{*\rm P} dt\, d\Gamma + \int_\Omega \int_{t_{F-1}}^{t_F} \left[\left(u v_j - a_{\rm N}\frac{\partial u}{\partial x_j}\right)\frac{\partial u^{*\rm P}}{\partial x_j} + \frac{I_u u^{*\rm P}}{\rho_o c_o}\right] dt\, d\Omega$$

$$+ \int_\Omega u_{F-1} u^{*\rm P}_{F-1} d\Omega \,. \tag{3.1.18}$$

3.2 Numerical Solution of Diffusion-Convective Equation

Searching for an approximate solution of unsteady diffusion-convective equation the corresponding boundary-domain integral representation is given in a discretised form in which the integrals over the boundary and domain are approximated by a sum of integrals over E individual boundary elements and C internal cells respectively, e.g. the discretised representation of the general nonlinear integral statement eq.(3.1.18) is

$$c(\xi)u(\xi, t_F) + a_o \sum_{e=1}^{E} \int_{\Gamma_e} \int_{t_{F-1}}^{t_F} u \frac{\partial u^{*\rm P}}{\partial n} dt\, d\Gamma = \sum_{e=1}^{E} \int_{\Gamma_e} \int_{t_{F-1}}^{t_F} a_e \frac{\partial u}{\partial n} u^{*\rm P} dt\, d\Gamma$$

$$
-\sum_{e=1}^{E}\int_{\Gamma_e}\int_{t_{F-1}}^{t_F} uv_n u^{*\mathrm{P}}\,dt\,d\Gamma+\sum_{c=1}^{C}\int_{\Omega_c}\int_{t_{F-1}}^{t_F}\left[\left(uv_j-a_{\mathrm{N}}\frac{\partial u}{\partial x_j}\right)\frac{\partial u^{*\mathrm{P}}}{\partial x_j}+\frac{I_u}{\rho_o c_o}u^{*\mathrm{P}}\right]dt\,d\Omega
$$

$$
+\sum_{c=1}^{C}\int_{\Omega_c} u_{F-1}u^{*\mathrm{P}}_{F-1}\,d\Omega\,. \tag{3.2.1}
$$

Notice, that for the constant material properties and linear source term, the above equation simplifies to a linear one. Next, the values of all field functions are to be approximated in each individual boundary element and internal cell by the use of interpolation polynomials $\{\Phi\}$ and $\{\phi\}$ with respect to space, and by time interpolation polynomials $\{\psi\}$. Let the index n refer to the number of nodes in each boundary element or internal cell, i.e. also relates to the degree of the respective spatial interpolation polynomials, while the index m means the degree of time polynomial $\{\psi\}$, e.g. $m = 1$ for constant, $m = 2$ for linear time fitting etc. With respect to the approximations used different solution procedures can be developed, e.g. the implicit-explicit and pure implicit numerical algorithm.

3.2.1 Iterative Implicit-Explicit Scheme

The following approximations of functions and products of functions in space and time are used

$$
\begin{aligned}
u &= \{\Phi\}^T\{\psi\}\{u\}^n_m \;, \\
a_e\frac{\partial u}{\partial n} &= \{\Phi\}^T\{\psi\}\left\{a_e\frac{\partial u}{\partial n}\right\}^n_m \;, \\
uv_n &= \{\Phi\}^T\{\psi\}\{uv_n\}^n_m \;, \\
uv_j-a_{\mathrm{N}}\frac{\partial u}{\partial x_j} &= \{\phi\}^T\{\psi\}\left\{uv_j-a_{\mathrm{N}}\frac{\partial u}{\partial x_j}\right\}^n_m \;, \\
\frac{I_u}{\rho_o c_o} &= \{\phi\}^T\{\psi\}\left\{\frac{I_u}{\rho_o c_o}\right\}^n_m \;, \\
u_{F-1} &= \{\phi\}^T\{u\}^n_{F-1} \;.
\end{aligned} \tag{3.2.2}
$$

Let us select the constant interpolation distribution of all field variables within individual time increment, e.i. $\psi = 1$. Time integrals in eq.(3.2.1) may be evaluated analytically for space and plane geometry cases

$$
\begin{aligned}
U^{*\mathrm{P}} &= a_o\int_{t_{F-1}}^{t_F} u^{*\mathrm{P}}\,dt \;, \\
\frac{\partial U^{*\mathrm{P}}}{\partial n} &= a_o\int_{t_{F-1}}^{t_F}\frac{\partial u^{*\mathrm{P}}}{\partial n}dt \;, \\
\frac{\partial U^{*\mathrm{P}}}{\partial x_j} &= a_o\int_{t_{F-1}}^{t_F}\frac{\partial u^{*\mathrm{P}}}{\partial x_j}dt \;.
\end{aligned} \tag{3.2.3}
$$

Next, in the discretised formulation eq.(3.2.1) are the following space integrals, which are functions of the geometry, time step τ and constant diffusivity a_o

$$h_e^n = \int_{\Gamma_e} \Phi^n \frac{\partial U^{*P}}{\partial n} d\Gamma \quad , \quad g_e^n = \int_{\Gamma_e} \Phi^n U^{*P} d\Gamma \ ,$$
$$d_{c_j} = \int_{\Omega} \phi^n \frac{\partial U^{*P}}{\partial x_j} d\Omega \quad , \quad d_c^n = \int_{\Omega_c} \phi^n U^{*P} d\Omega \ , \tag{3.2.4}$$
$$b_c^n = \int_{\Omega_c} \phi^n u^*_{F-1} d\Omega \ .$$

With these integrals the discretised equation can be written in the following form

$$c(\xi)u(\xi,t_F) + \sum_{e=1}^{E} \{h\}^T \{u\}_F^n = \sum_{e=1}^{E} \{g\}^T \left\{\frac{a_e}{a_o}\frac{\partial u}{\partial n}\right\}_F^n - \frac{1}{a_o}\sum_{e=1}^{E} \{g\}^T \{uv_n\}_F^n$$
$$+ \frac{1}{a_o}\sum_{c=1}^{C} \{d_j\}^T \left\{uv_j - a_N \frac{\partial u}{\partial x_j}\right\}_F^n + \frac{1}{a_o}\sum_{c=1}^{C} \{d\}^T \left\{\frac{I_u}{\rho_o c_o}\right\}_F^n + \sum_{c=1}^{C} \{b\}^T \{u\}_{F-1}^n \ . \tag{3.2.5}$$

The discretised statement eq.(3.2.5) is formally identical to steady diffusion-convective discretised formulation eq. (2.2.5), thus basicaly the same numerical solution procedure can be applied to obtain the approximate numerical solution of the unsteady case for each individual time increment. The equation has to be first written for all boundary nodes using collocation method, $\xi = 1, N_e$, where N_e indicates the number of all boundary nodes, rendering an implicit system of equations for the boundary unknowns only, $\{u\}_F$ or $\{a_e/a_o\ \partial u/\partial n\}_F$ respectively, e.g. given in a matrix notation form

$$[c(\xi)]\{u(\xi,t_F)\} + \left[\widehat{H}\right]\{u\}_F = [G]\left\{\frac{a_e}{a_o}\frac{\partial u}{\partial n}\right\}_F - \frac{1}{a_o}[G]\{uv_n\}_F$$
$$+ \frac{1}{a_o}[D_j]\left\{uv_j - a_N\frac{\partial u}{\partial x_j}\right\}_F + \frac{1}{a_o}[D]\left\{\frac{I_u}{\rho_o c_o}\right\}_F + [B]\{u\}_{F-1} \ , \tag{3.2.6}$$

or

$$[H]\{u\}_F = [G]\left\{\frac{a_e}{a_o}\frac{\partial u}{\partial n}\right\}_F - \frac{1}{a_o}[G]\{uv_n\}_F$$
$$+ \frac{1}{a_o}[D_j]\left\{uv_j - a_N\frac{\partial u}{\partial x_j}\right\}_F + \frac{1}{a_o}[D]\left\{\frac{I_u}{\rho_o c_o}\right\}_F + [B]\{u\}_{F-1} \ , \tag{3.2.7}$$

where the diagonal matrix $[c(\xi)]$ is added to the matrix $[\widehat{H}]$.

References

[1] Backer A.J.: *Finite Element Computational Fluid Mechanics,* Hemisphere Publ. Corp., NY, 1983

[2] Banerjee P.K., Butterfield R.: *Boundary Element Techniques in Engineering Science.* Mc.Graw-Hill, NY 1981

[3] Brebbia C.A.: *The Boundary Element Method for Engineers.* Pentech Press, London, Halstead Press, NY, 1978

[4] Brebbia C.A., Telles J.C.F., Wrobel L.C.: *Boundary Element Methods - Theory and Applications.* Springer Verlag, NY, 1984

[5] Brebbia C.A., Skerget P.: *Time Dependent Diffusion-Convection Problems Using Boundary Elements.* 3rd Int.Conf. on Num.Meth in Lam.and Turb. Flow, Seatle, 1983

[6] Hebeker F.K.: *Characteristics and Boundary Elements for Navier-Stokes Flows.* 9th. Int. Conf. on BEM, Stuttgart. Springer Verlag, Berlin, 1987

[7] Ikeuchi M., Sakakihara M. and Onishi K.: *Constant Boundary Element Solution for Steday-State Convective Diffusion Equation in Three Dimensions.* Transactions of the IECE of Japan, 1983

[8] Onishi K., Kuroki T., Tanaka M.: *Boundary Element Method for Laminar Viscous Flow and Convective Diffusion Problems.* Ch.8, Topics in Boundary Element Research, V.2 pp.209-229, Springer Verlag, Berlin, 1985

[9] Rizk Y.M.: *An Integral Representation Approach for Time Dependent Viscous Flows.* PhD Thesis, Georgia Institute of Technology, Atlanta, 1980

[10] Roache P.J.: *Computational Fluid Dynamics.* Hermosa Publ., 1982

[11] Skerget P., Zagar I., Alujevic A.: *Three-Dimensional Stady State Diffusion-Convection.* International conference on Boundary Elements,9., Vol.3: Fluid flow and potentional applications, Stuttgart 1987

[12] Skerget P., Brebbia C.A.: *The Solution of Convective Problemsin Laminar Flow.* 5th Int. Conf.on BEM, Hiroshima. Springer Verlag, Berlin, 1983

[13] Skerget P., Alujevic A., Brebbia C.A.: *The Solution of Navier-Stokes Equations in Terms of Vorticity-Velocity Variables by Boundary Elements.* 6th Int.Conf. on BEM, Southampton. Springer Verlag, Berlin, 1984

[14] Skerget P., Alujevic A., Brebbia C.A.: *Analysis of Laminar Flows With Separation Using BEM.* 7th Int. Conf. on BEM, Como. Springer Verlag, Berlin, 1985

[15] Skerget P., Alujevic A., Brebbia C.A.: *Analysis of Laminar Fluid Flows by Boundary Elements.* 4th. Int. Conf. on Num. Meth. in Lam. and Turb. Flows, Swansea, Pineridge Press, Swansea,1985

[16] Skerget P., Alujevic A., Brebbia C.A.: *BEM for Laminar Motion of Isochoric Viscous Fluid.* 2nd BETECH Conf. Cambridge/Mass. Springer Verlag, Berlin, 1986

[17] Skerget P., Alujevic A., Brebbia C.A.: *Nonlinear Transport Problems in Fluids by Boundary Elements.* 3rd Int. Conf. on Num. Meth. in Nonlinear Problems, Dubrovnik. Pineridge Press, Swansea, 1986

[18] Skerget P., Alujevic A., Brebbia C.A. Kuhn G.: *Boundary Elements for Mixed Convection Problems.* 3rd Int. Conf. on BETECH, Rio de Janiero. Springer Verlag Berlin,1987

[19] Skerget P., Kuhn G., Alujevic A.: *Analyisis of Mixed Convection Flow Problems by BEM.* Zangew. Math.Mech., Vol.68, 1988

[20] Skerget P., Alujevic A., Kuhn G., Brebbia C.A.: *Natural Convection Flow Problems by BEM.* 9th Int. Conf. on BEM, Stuttgart. Springer Verlag, Berlin, 1987

[21] Skerget P., Alujevic A., Zagar I., Brebbia C.A., Kuhn G.: *Time Dependent Three-Dimensional Laminar Isohoric Viscous Fluid Flow by the Boundary Element Method.* Proc. 10th Int. Conf. on Boundary Elements, Vol.2, pp.57-74. Springer Verlag, Berlin 1988

[22] Smith A.M.O., Pierce J.:*Exact Solution of the Neumann Problem. Calculation of Plane and Axially Symmetric Flow About or Within Arbitrary Boundaries.* Proc. 3rd US National Congress of Appl. Mech., pp.807-815, 1958

[23] Tanaka Y., Homma T. and Kaji I.: *Mixed Boundary Element Solution Using Both of Constant and Linear Elements for a Steady-State Convective Diffusion Problem in Three Dimensions.* Transactions of the IECE of Japan, 1984

[24] Taylor C., Hughes T.G.: *Finite Element Programming of the Navier-Stokes Equations.* Pineridge Press, Swansea, 1981

[25] Tosaka N., Kakuda K.: *Numerical Simulation for Incompressible Viscous Flow Problems Using the Integral Equation Methods.* 8th Int.Conf. on BEM, Tokyo. Springer Verlag, Berlin, 1986

[26] Tosaka N., Fukushima N.: *Integral Equation Analysis of Laminar Convection Problems.* 8th Int. Conf. on BEM Tokyo. Springer Verlag, Berlin, 1986

[27] Tosaka N., Kakuda K.: *Numerical Simulations of Laminar and Turbulend Flows by Using an Integral Equations,* 9th Int.Conf. on BEM, Stuttgart. Springer Verlag, Berlin, 1987

[28] Wahbah M.M.: *Computation of Internal Flows With Arbitrary Boundaries Using the Integral Representation Method.* School of Aerospace Engineering. Georgia Inst. of Technology, Atlanta (DAAG 29-75-G-0147), 1975

[29] Wu J.C., Thompson J.F.: *Numerical Solution of Time Dependent Incompressible Navier-Stokes Equations Using an Integro-Differentional Formulation.* Comp. and Fluids, Vol.1, pp.197-215, 1973

[30] Wu J.C.: *Problems of General Viscous Flow.* Developments in BEM, V.2, Ch.2. Elsevier Appl.Sci.Publ., London, 1982

[31] Wu J.C., Rizk Y.M., Sankar N.L.: *Problems of Time Dependent Navier-Stokes Flow,* Developments in BEM, V.3., Ch.6., Elsevier Appl.Sci.Publ., London, 1984

[32] Zagar,I., Skerget,P., Alujevic, A.: *Pressure Drop Calculations for Tube Bundles in Crossflow by the Boundary-Domain Integral Method.* Zeitschrift für Angewandte Mathematik und Mechanik, 72 (1992) 5.

[33] Zagar,I., Skerget,P., Alujevic, A.:*Boundary-Domain Integral Method for the Space Time Dependent Viscous Incompressible Flow.* Boundary Integral Methods, Theory and Applications (ed. Morino, L. and Piva, R.), Proceedings of the IABEM Symposium Rome, Italy, October 15-19, 1990

AUTHORS' INDEX

LUCY CAVENDISH COLLEGE